T0258469

Feverish Bodies, Enlightened Minds

Feverish Bodies, Enlightened Minds

SCIENCE AND THE YELLOW FEVER CONTROVERSY IN THE EARLY AMERICAN REPUBLIC

Thomas A. Apel

STANFORD UNIVERSITY PRESS
STANFORD, CALIFORNIA

Stanford University Press
Stanford, California

Printed in the United States of America on acid-free, archival-quality paper

Library of Congress Cataloging-in-Publication Data

Names: Apel, Thomas A., author.
Title: Feverish bodies, enlightened minds : science and the yellow fever controversy in the early American republic / Thomas A. Apel.
Description: Stanford, California : Stanford University Press, 2016. | (c)2016 | Includes bibliographical references and index.
Identifiers: LCCN 2015041710 (print) | LCCN 2015042805 (ebook) | ISBN 9780804797405 (cloth : alk. paper) | ISBN 9780804799638 (ebook)
Subjects: LCSH: Yellow fever—United States—History—18th century. | Yellow fever—Etiology—History—18th century. | Epidemics—United States—History—18th century. | Diseases—United States—History—18th century. | Medical sciences—United States—History—18th century. | Diseases—Causes and theories of causation—History—18th century.
Classification: LCC RA644.Y4 A64 2016 (print) | LCC RA644.Y4 (ebook) | DDC 614.5/41—dc23
LC record available at http://lccn.loc.gov/2015041710

Typeset by Newgen in 11/14 Adobe Garamond

To Amanda,
for her love and patience

Contents

Acknowledgments

A number of people and institutions helped make this book possible, and I would like to thank as many of them as I can recall. First, the libraries and archives where so much of the research and writing of this book happened. The Georgetown University Library and its staff were always warm, receptive, and helpful. I also want to thank the staff at the Milton S. Eisenhower Library at the Johns Hopkins University, my home away from home, where I did the lion's share of the reading and writing for this book. The libraries of UC Berkeley, the Library Company of Philadelphia, the Historical Society of Pennsylvania, the Historical Society of New York, the New York Academy of Medicine Library, the Maryland Historical Society, the Library of Congress, the National Library of Medicine, and the Columbia University Archives were all instrumental in the research for this book. Finally, I am indebted to the Huntington Library and the Jack Miller Center for a generous and productive fellowship spent during the summer of 2013 in beautiful San Marino, California.

I have been lucky to know a number of insightful scholars who have provided input and given feedback on parts of this book: Kathryn Olesko, Alison Games, Michael Kazin, James Collins, Josiah Osgood, Jessica Simmon Hower, Michael Hill, and Darcy Kern. My scholarship truly came of age with the help of John R. McNeill, who provided a model of scholarly excellence and originality, and my friend Adam Rothman, who was there from the time this project formed as an inchoate bundle of ideas, to the point where it took the form of a book. I also want to thank Kyle Roberts for his friendly advice over the years. Jan Golinski provided helpful commentary on Chapter Three of this book, and David Waldstreicher helped immensely with Chapter Two. Thank you as well to the anonymous reviewers of this book, whose comments helped sharpen its thrust. I sincerely appreciate Eric

Brandt, Friederike Sundaram, and the entire editorial staff at the Stanford University Press.

Finally, I want to thank my family. My parents instilled in me a love of books and learning, and they supported me when I conceived the notion of turning that love into a career. My brother and sister, Mickey and Hillary, have been companions and confidants since my earliest days. My wife, Amanda Reider, a scientist and mother, among many other roles, has helped me in ways that I cannot tally up or describe. And much love to my daughter, Ainsley Ione Apel, whose birth in 2014 makes this manuscript the second best thing I produced that year.

Feverish Bodies, Enlightened Minds

Introduction

In 1793, pestilence visited Philadelphia, the United States' political capital and its center of economic, cultural, and scientific activity. Doctors quickly identified the disease. Referring to it alternately as the "bilious remitting" fever, the "malignant" fever, or even the "synochus icteroides," they all nevertheless recognized it by its more common name: yellow fever. They hardly could have mistaken it. Yellow fever strikes in two distinct phases. In the first, victims exhibit fever, headache, chills, languor, and in certain cases nausea and vomiting. Patients then experience a remission, from which most emerge unscathed and without relapse. Those unfortunate enough to experience the second stage suffer an intensification of the fever, delirium, jaundice (caused by damage to the liver, which usually constitutes the final cause of death), and, finally, the dreaded "black vomit," a foul mixture containing partially digested blood and an almost sure sign of approaching death. In modern settings, perhaps one in ten victims of yellow fever will die, but in the Philadelphia of 1793, a city of about 50,000 people, the disease exacted an even greater toll.[1] Appearing first in late July, along the bustling wharves of the commercial center, yellow fever soon infiltrated the adjoining neighborhoods, and then the city as a whole. The fever raged for the rest of August, September, and October, until finally the scourge ceased with the frosts of the approaching winter. All told, the epidemic of 1793 carried off as many as 5,000 lives in the short span of three months.[2]

The story of Philadelphia's great plague has been told many times before, but few emphasized the full extent of the yellow fever problem.[3] Far from disappearing after 1793, the disease returned to Philadelphia in 1797 (about 1,500 dead), 1798 (3,645 dead), 1799 (about 1,000 dead), and then less severely in 1802, 1803, and 1805. New York hosted yellow fever almost every year from 1795 to 1805, with major epidemics in 1795 (800 dead), 1798 (2,080 dead), and 1803 (about 700 dead). Baltimore endured a major epidemic in 1800 (1,197 dead), and it, along with Boston, Charleston, and New Orleans, hosted several relatively minor epidemics, which each nevertheless resulted in dozens, sometimes hundreds, of deaths each.[4]

Yellow fever constituted the most pressing natural problem of the early national period. Besides the deaths, yellow fever incited frantic, mass evacuations. It halted commerce in the nation's busiest port cities for months at a time and led to burdensome quarantines and expensive sanitary reform measures. On several occasions, the disease also interfered with the workings of the national government, forcing the president and Congress to evacuate Philadelphia, and leading ultimately to the removal of the capital to the Potomac. More alarmingly still, yellow fever eroded public virtue, the cornerstone of a healthy republic. In his popular *Short Account of the Malignant Fever Lately Prevalent in Philadelphia* (1793), the Irish printer Mathew Carey horrified readers with scenes of familial betrayal and ruthless self-interest, concluding that the fever produced "a total dissolution of the bonds of society in the nearest and dearest connexions."[5] The crisis of yellow fever appeared to involve the fate of the republic. Writing years later, a young doctor named Stubbins Ffirth predicted that, if left unchecked, yellow fever would lead to the "loss of our commerce by shutting all foreign ports against our vessels, and of course the annihilation of our agriculture, our manufactures, and the down fall of the fair superstructure of science and of liberty."[6]

With the future of the republic and the lives of its citizens hanging in the balance, leading medical and scientific[7] thinkers hoping to prevent the disease posed a deceptively simple question: What caused yellow fever? The search for the cause of yellow fever involved many prominent American intellectuals, such as Benjamin Rush and Noah Webster, and provoked a heated and divisive debate between "localists," who believed that the disease arose from locally situated miasmas, and "contagionists," who thought it came from abroad and could be transmitted from person to person. Adherents of both theories offered compelling rationales. Arguing that it arose from domestic sources, localists noted that the disease only occurred during the hottest and wettest periods of the year, and that it only prevailed in the confines of cities, never beyond. Contagionists countered that yellow fever outbreaks always coincided with the arrival of infected vessels from the West Indies, where the disease commonly occurred. In most instances, they could even trace the first cases of yellow fever to individuals from disease-ridden ships. The polarity of the debate reflected the deceptive etiology of the disease. Yellow fever is caused by a virus transmitted through the bites of *Aedes aegypti* mosquitoes. Finicky animals, *A. aegypti* prefer to breed in small artificial containers of water, especially those that abound in cities, and they die when the temperature drops below about 43°F (hence the localists' insistence on the climatic and urban specificity of the disease). Since the mosquitoes die in the frosty winters of the northeastern United States, both the virus and the vector had to be reintroduced each year in order for outbreaks to occur (hence the contagionists' implication of ship traffic).[8]

The seeming intractability of yellow fever debate, no less than the urgency of the disease itself, tested and strained the fever investigators, altering as it did the nature of the problem. The crisis pushed them to explore fields of inquiry not typically associated with disease thought. Some composed massive histories of disease, hoping that the patterns of disease outbreaks would lead to some understanding of the cause of yellow fever. Others embraced the chemical revelations associated with Antoine-Laurent Lavoisier and examined the chemical makeup of the matter that caused yellow fever. Still others pondered the natural theology of yellow fever—its purpose as it appeared from the evidence of design. By the end of the epidemic period, investigators of yellow fever had produced one of the most substantial and innovative, yet underexplored, outpourings of scientific thought in American history. In each area of study, the localists pushed the debate in their favor. By the end of the epidemic period, they had more or less settled the debate, and the contagionists either converted or retired from public view.

Still, though the localists effectively won the debate in their own time, they did not determine the true cause of yellow fever. That feat came about a hundred years later, when the United States Army Yellow Fever Commission led by Walter Reed, investigating the theory of the Cuban doctor Carlos Finlay, performed the experiments that identified *A. aegypti* as the vector of yellow fever. Another quarter century later, researchers located the virus responsible for the disease. Both originated in Africa thousands of years ago and were introduced to the New World in the seventeenth century as part of the Columbian Exchange.[9] From 1793 to 1805, rapid population growth in American port cities, combined with warfare in the West Indies, provided the necessary epidemiological ingredients for pandemic yellow fever. By prompting the introduction of thousands of nonimmune soldiers, the Haitian Revolution supplied female *A. aegypti* mosquitoes with plenty of vulnerable humans to feed on and infect with the yellow fever virus. With the vector and virus flourishing in the martial environment, merchants, soldiers, and refugees easily carried them to the United States, only a short distance away.[10]

All this begs the question: If the investigators failed to determine the true cause of yellow fever, and if they only supported theories that are now dead and almost forgotten, why study their efforts? For generations, historians more concerned with tales of progress did not. A host of perspectives coming mainly from European history offer ways of evaluating and appreciating even the most seemingly outdated scientific views. These works have called attention to the way that knowledge is "constructed," and the way that deconstructing knowledge yields insights about people and societies, their overarching philosophical and scientific notions, and the multifaceted historical contexts that influenced their understandings of the natural world. As Thomas Kuhn wrote of the matter, "An apparently arbitrary element, compounded of personal and historical accident, is always a formative ingredient of the beliefs espoused by a given scientific community at a given time."[11] A growing body of scholarship on the disease in the United States has already begun to explain yellow fever's relevance to a broader range of issues, from slavery and commerce, to public health and the public sphere.[12] But the yellow fever controversy still lacks a book-length study that considers it for what it ultimately was—a debate about nature and science, one that both exposed the foundations of early American scientific knowledge production and undermined them.

The yellow fever ferment opens an ideal window onto the contours of natural inquiry in the early republic. Because of its mysteriousness, yellow fever explicitly forced investigators to probe the limits of their knowledge of nature, its capacities, and its design. It also mobilized more inquirers, and it inspired them to reach out to a more diverse range of fields, than any other scientific problem of the era. Yellow fever inquiry reveals that the domains of knowledge that gathered under the umbrella of science were broader than historians have acknowledged, and that these domains of knowledge lacked clear-cut boundaries.[13] Because of its divisiveness, the yellow fever controversy exposes the conduct of natural inquiry—the ways natural philosophers communicated with themselves and the public, organized themselves into discursive communities, and offered and contested knowledge claims. Finally, because the localists settled the debate, the yellow fever controversy reveals the standards upon which accepted knowledge rested. It promises, in other words, to reveal the modes of early republican knowledge production, and to identify what arbitrary historical elements shaped the victory of the localists (despite having no more "facts" to support their theory).[14]

Explaining the localist supremacy resurrects a familiar problem for historians who have studied disease thought in nineteenth-century Europe, where the battle between localists and contagionists raged on (in the search for the cause of cholera in particular) and localists held sway until the microbiological breakthroughs of the 1860s. Following Erwin Ackerknecht, many have argued that the uneven allegiances to competing causal theories fell along political lines. As Ackerknecht argued, localism appealed to liberal, commercially oriented Western Europeans, who wanted to liberate commerce from the shackles of quarantines, which always came with contagionism, and free the individual from the encroachments of the state. Contagionism, by contrast, persisted in the autocratic countries of Eastern and Central Europe, where such sentiments did not flourish. One historian has even attempted to correlate causal theories with political allegiances in 1793 Philadelphia—statist, pro-urban, Francophobic federalists blamed the disease on the influx of French refugees from Saint-Domingue and Europe, whereas the liberal, anti-urban, Francophilic republicans embraced localism.[15]

Ackerknecht's thesis and the works it has inspired rightly call attention to the political and social contexts that shaped disease inquiry, but they also simplify the factors that differentiated the opposing sides and privilege public health. They cast prevailing health policies as the ultimate ends of all inquiry as well as indexes of popular opinions about disease causation, and they focus on the thoughts, ideas, and procedures that emerged from public health bodies themselves. When yellow fever first struck the United States in 1793, however, public health structures capable of imposing quarantines or enforcing sanitary reform were almost nonexistent. Inquiry into the cause took place in public, not behind the closed doors of public health bodies. Yellow fever precipitated comprehensive public health reform, but measures never favored one causal theory or the other. Cities such as New York and Philadelphia erected rigorous quarantine procedures and thoroughgoing sanitation measures, including waterworks systems.[16] At least for these investigators, the increasingly urgent tone of the debate did not reflect anxieties about public health alone. They also labored to define the parameters of acceptable natural inquiry, and to impose their definitions on what counted as knowledge.

Far from splitting along political lines, investigators actually shared more similarities than differences. Well-educated, white males, they were almost all devoutly religious, and they enjoyed at least a modicum of wealth, leisure, and social standing. They cast themselves as gentlemen and natural philosophers, and they stressed their commitments to plain "facts" and the accumulation of useful knowledge. Most, though not all, were trained doctors. The yellow fever conundrum more properly qualified as a problem of science, not medicine. As Noah Webster wrote, "I . . . consider the question as resting principally on fact, and not on medical skill; therefore proper to be investigated and discussed by any man who has leisure and means, as well as by physicians."[17] Besides, most early republicans embraced the ideal of open and egalitarian public discourse, and so viewed claims to authority with suspicion. Truth claims rested not on the authority of the claimant, but upon the approval of the public.[18]

Only upon closer examination of the investigators do their deeper philosophical differences come into view. Localists from the Jeffersonian-republicans Benjamin Rush and Samuel Latham Mitchill to the arch-federalist Noah Webster studied at Scottish universities, or those modeled after them in the United States. They embraced the essential precepts of the latest scientific methodologies—knowledge of nature was to be attained inductively, through the slow accumulation of empirical facts and data. As pious Protestants, they also believed the natural, its laws and its wonder, revealed the will and design of God. And like the Scots from whom they adapted the concept, localists believed that God had endowed human beings with an innate rational capacity known as common sense. A concept devised to free inquirers from crushing skepticism and validate self-evident knowledge (sensory knowledge and matters of fact), common sense became in the hands of localists a malleable tool that allowed them to grasp the underlying logic of the world, to unravel its mysteries, and to find truth. Common sense had a two-fold reality—it was at once the perceiving organ of the mind itself, and it was the obvious, common-sense truths it detected in the world.[19] It exerted a powerful influence over the yellow fever inquiry.

Contagionists were no less pious or scientific, and perhaps no less philosophical than the localists, but they refused to let the same considerations enter into their inquiries. Taught through apprenticeships, in hospitals and with hands-on experience, they tended toward materialism, evidenced by a predilection for surgery (a mere technical skill, according to university-trained physicians, and one they looked on condescendingly) and experimental chemistry. They consistently criticized the localists for their speculations and flights of fancy, and they adhered more strictly to the facts of yellow fever. Contagionists even sometimes admitted that local environmental conditions might activate particles of yellow fever into epidemic proportions. They focused principally on tracing and intercepting the pathways of the contagious particles, which they thought were ultimately responsible for yellow fever. For them, no arguments, no rationales, no elaborate justifications negated the convincing and well-founded fact that infected vessels arrived before every outbreak, and that the first victims came from those pestilential vehicles or nearby.

Why, then, did the contagionists' convincing and well-founded arguments fail to hold public opinion, and why did the localist perspective prevail? One factor was purely practical. Contagionists did not publish as voraciously as localists, nor did they did occupy university positions from which they perpetuated their ideas to scores of young students. On the whole, localists made themselves more visible in the public sphere, and they better organized themselves into a discursive community that privately coordinated what they put forward publicly. I will declare from the outset that because they produced so much more than contagionists, and with so much more imagination, the localists garner far more attention in this book. Regrettably there are moments when the contagionists all but fall out of the picture, but this is an unavoidable product of the nature of the yellow fever debate and not an arbitrary decision of the author.

More importantly, localist explanations better appealed to common sense. Localists emerged victorious, not simply because they shouted more loudly and certainly not because they marshaled better facts than the contagionists, but because they more plausibly depicted yellow fever as an entity that conformed to the apparent design of the world, one that could be understood with common faculties of the mind. They persuaded audiences that locally generated diseases occurred in cycles that appeared throughout history; they rationalized the chemical construction of the matter that brought on yellow fever, and explained its emergence from normal chemical reactions; and, finally, they used evidence of design and scripture to explain why it would exist in God's world, and how people could stop it. In a religious and enlightened society, where a learned public followed the debate with great interest and anxiety, the *plausibility* of localism outweighed the facts of contagionism.

The triumph of localism showcases the formative influence of religion in early republican science. This is not a new realization, to be sure. Many have emphasized the religious orientation of American scientific fields and philosophies. But again, because investigators subjected yellow fever to such a broad range of scientific analysis, the yellow fever controversy uniquely shows how religious ideas cut across fields and crystallized in the notion of common sense.[20] In the minds of natural philosophers from Rush to Mitchill, truth of nature did not descend from observation and induction alone; it also arose from rational, common-sense reflection on the apparent design of the world, whereby the proof of an idea rested on its coherence and understandability. We should abandon the notion that the arch-empiricist Thomas Jefferson—"I feel: therefore I exist," he claimed—best represents American scientific practice in the early republic. Science on a common-sense model, more than a strict empirical or experimental one, suited the social and intellectual landscape of the early republican United States, and provided an overarching paradigm for natural inquiry.[21]

Religion and its philosophical manifestation as common sense helped the localists win the debate, but it was only one of the "arbitrary historical" elements that directed the yellow fever contest. Ideas about yellow fever did not rest on simple, disconnected contemplations of philosophical or religious principles any more than the facts of the disease. Ackerknecht and those who came after were correct to emphasize the broader political dimensions of disease thought. The debate took shape in the crucible of the 1790s, an age of revolutionary upheaval in the greater Atlantic world. Yellow fever informed opinions about the direction of the republic and the world. Examining the history of disease, for example, forced early republicans to reconsider their hopes of escaping history and constructing a *novus ordo seclorum*, and it contributed to their fears of cities. Contemporary events, such as the French Revolution and the diffusion of deistic and atheistic thought, also influenced the way that investigators, especially localists, thought about yellow fever. These threats mediated localists' reception of Lavoisier's chemistry, inspired the religious zeal with which they promoted their public health remedies, and impelled them to articulate and defend a philosophical system that rescued human inquiry from heretical skepticism.

The yellow fever debate shared a deeper relationship with contemporary political discourse. In both, participants accused their opponents of joining factions and intentionally distorting the truth. These parallel discourses reveal structural similarities between disease thought and politics: both aspired to the status of sciences and seemed to appeal to common-sense considerations, and both were rooted in the same material organization. They also proved mutually reinforcing—one aided the believability of the other—and provoked far-reaching backlashes. The acrimony of the yellow fever debate fractured the intellectual community of natural inquirers and left investigators wanting to exert greater top-down control over the course of natural inquiry, just as the bitterness of the 1790s political wars left intellectuals wanting to contain and channel political discourse. The yellow fever years thus help explain the trajectory of American science and medicine in the nineteenth century, when these forms of inquiry moved from the public sphere into professional organizations and institutions of "experts," where they became objects of greater disciplinary and methodological rigor.[22] The yellow fever debate also points to new ways of considering changes in early republican public discourse.

What follows is a history of the yellow fever years, the people who studied the dreadful disease, and above all the ideas that they brought to their investigations. Structurally, *Feverish Bodies, Enlightened Minds* proceeds thematically, with chapters devoted to each area of inquiry the investigators brought to bear on the problem of yellow fever. Chapter 1 frames the parameters of the debate, establishes the intellectual orientations of the investigators, and highlights in particular their interests in the "facts" of yellow fever. Chapter 2 discusses the investigators' uses of historical sources and their compositions of disease histories in their efforts to evaluate the patterns through which diseases such as yellow fever normally operated, and thereby to shed light on the origins of the yellow fever epidemics. Chapter 3 follows the investigators' appropriation of Antoine-Laurent Lavoisier's groundbreaking discoveries in chemistry, collectively dubbed the "Chemical Revolution," in order to apprehend the chemical construction of the invisible particles that caused yellow fever. Chapter 4 deals more directly with the investigators' religious understandings of yellow fever, and the localists' explicit efforts to reconcile yellow fever with God's purpose and to invest their public health solutions with a crusader's zeal. Chapter 5 examines the conspiratorial tone of the debate, and the way that the fear of factions fractured the intellectual community of natural inquirers and left investigators wanting to exert greater top-down control over the course of natural inquiry.

We may now return to Philadelphia in 1793, when the disease made its first appearance and the problem of yellow fever took form.

Contexts and Causes

> We [doctors] are employed in . . . a necessary calling
> that enforces to us the weakness and mortality of
> human nature. This earthly frame, a minute fabric, a
> center of wonders, is forever subject to Diseases and
> Death. The very air we breathe too often proves
> noxious, our food often is armed with poison, the very
> elements conspire the ruin of our constitutions, and
> Death forever lies lurking to deceive us.[1]
>
> BENJAMIN RUSH, *1761*

What caused yellow fever? The question was on most everyone's mind in the United States' capital in the winter of 1793, as legions of Philadelphians, some twenty thousand in all, trudged back into the city they had abandoned only months before. Those seeking answers would have found them, lots of them; indeed, more than they would have liked. People were talking and the newspapers were all abuzz with word of the disease. According to Mathew Carey, "almost all" Philadelphians believed that the disease came from abroad, probably along with the two thousand French-speaking refugees from Saint-Domingue who had arrived in the ports of Philadelphia earlier that summer, some of them allegedly infected with yellow fever.[2] Others implicated the conditions of the city: the stinking cesspools, the decaying animal carcasses, rotten vegetables, and stagnant pools of water that bred pestilential miasma. A pseudonymous writer for John Fenno's *Gazette of the United States*, masquerading as William Penn, the founder of Philadelphia writing from the "*Elysian fields*," chided his contemporaries for straying from his plan for the city. "It was my intention in laying the plan of Philadelphia," the ghost of Penn wrote, "to provide for the *health* as well as the beauty and conveniency of the place" (Figure 1.1). The author, evidently a localist, blamed city-dwellers for cutting down trees, narrowing the streets, and building where they should not. "Had you preserved my plan," the imposter informed his readers, "you might have avoided the mischief which has now befallen you."[3]

Those learned in science and medicine set out to solve the puzzle. Benjamin Rush opened the commentary. The forty-eight-year-old professor of the theory and practice of medicine at the University of Pennsylvania was one of the foremost physicians in the United States, and an established figure

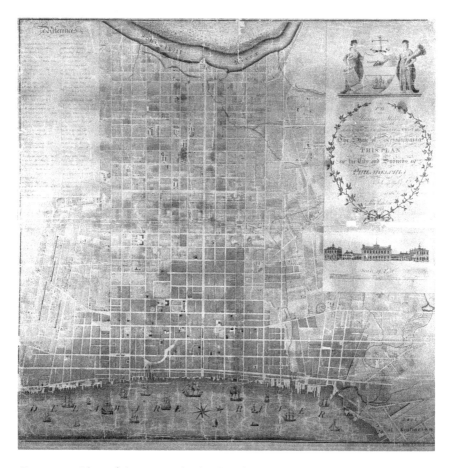

Figure 1.1.   Plan of the city and suburbs of Philadelphia, 1794.
SOURCE: Historical Society of Pennsylvania Of 610 1794.

in the nation's intellectual life. Born in Byberry, Pennsylvania, on Christmas Eve 1745, Rush studied medicine at the University of Edinburgh under the illustrious William Cullen. Once back in Philadelphia, Rush became the professor of chemistry at the College of Philadelphia, and he joined the American Philosophical Society. He later served in the Continental Congress and signed the Declaration of Independence. Rush labored hard to improve the medical and scientific reputation of Philadelphia, a city he rather hopefully deemed the *"Edinburgh of America."*[4] To that end, he helped form the College of Physicians of Philadelphia in 1789. A fierce Protestant, full of evangelical fervor, Rush also championed numerous causes—he campaigned for temperance, wrote essays against slavery, pushed for prison reform and women's rights, and argued for municipal sanitary reform. But there was also a darker side to the good doctor, a domineering intellectual who did not suffer contrary opinions.[5]

No one would have been surprised when Rush's localist manifesto, *An Enquiry into the Origin of the Late Epidemic Fever in Philadelphia*, began appearing at the city's numerous bookshops in December. Yet for all his erudition, authority, and gravitas, Philadelphia was scene to a vibrant and disputatious public culture, and so Rush immediately encountered opposition. Late in 1793, William Currie published a pointed rebuttal of Rush's first treatise. The son of an Episcopal clergyman, Currie like Rush was a native of Pennsylvania, and a member of both the American Philosophical Society and the College of Physicians. But Currie differed from Rush in a more pertinent respect—he lacked formal medical education. Though he attended some courses at the College of Philadelphia, he learned his trade chiefly through an apprenticeship with Dr. John Kearsley and as a surgeon for the Continental Army in the American Revolutionary War.[6] Currie's humble origins, his hands-on training, and his proficiency in the manual skill of surgery contrasted with Rush's impeccable education credentials and his more philosophical approach to medicine. The rivalry that emerged in the aftermath of the first outbreak lasted for the rest of the epidemic period.

And so, in the spring of 1794, when Rush published *An Account of the Bilious Remitting Yellow Fever*, a far lengthier effort, Currie immediately followed with his own full-length work, *A Treatise on the Synochus Icteroides, or Yellow Fever*. By that time, too, several others had joined the conversation. In their own treatises, Jean Devèze, an emigrant doctor from Saint-Domingue, and David de Isaac Cohen Nassy, a doctor and member of the

American Philosophical Society, sided with the localists. On the other side, Dr. John Beale Bordley Jr., the scion of a wealthy Maryland family, and Isaac Cathrall, a doctor and surgeon, decided on importation. Even the bookseller, Carey, entered the fray with a short contagionist essay.

Epidemics in Baltimore in 1794 and 1795, and New York in 1795 and 1796, elicited similar responses.[7] Yet, as the winter of 1796–97 settled over the Eastern Seaboard, localists and contagionists still vied for recognition as the purveyors of the authoritative explanation for the cause of yellow fever. The stalemate calls for examination of the investigators' ideas about disease causation, and about nature and the proper means of investigating it. Captivated by science and its grounding in empirical evidence and inductive reasoning, the investigators sought the facts of yellow fever. But since the facts reflected the etiology of yellow fever, the facts only deepened the controversy. One group sought answers from more elusive sources. Weaned on Scottish philosophy and taught that God's goodness suffused everything in creation, the localists insisted that the matter could be decided by common sense, a mental faculty that reveals as much about localist preconceptions of nature as the broader ideational forces hovering around natural inquiry in the late eighteenth century. After the first few years, the investigators remained bitterly opposed, but they had exposed the philosophical flaws that invested the problem of yellow fever with new significance and drove the debate forward.

In the late eighteenth century, Western-educated medical thinkers accepted localist and contagionist models of disease transmission. Both had ancient lineages, though contagionism most certainly predated its counterpart. Evidence from ethnographic fieldwork and early historical texts indicates that people predictably envision disease as an entity that can be caught from others. At the least, humans across time and space metaphorically construe even nontransmissible diseases as contagious and shun their victims as tainted. When disease struck Athens during the Peloponnesian War, citizens predictably concluded that it arrived from Ethiopia. In biblical texts, the fear of contagion translated into elaborate laws and procedures for isolating the sick and purifying infected areas.[8]

Contagion remained a vague concept until the seminal work of the Veronese humanist scholar Girolamo Fracastoro (1478–1553). In a groundbreaking work, *De Contagione et Contagiosis Morbis et Eorum Curatione*

(1546), Fracastoro offered a comprehensive view of contagions as specific types of "imperceptible particles," or *seminaria*. According to him, once *seminaria* implanted in a human body, the infected body produced identical *seminaria*, which could then be transmitted to others. The *seminaria* possessed unchanging natures and they could be reproduced infinitely, so long as they came into contact with bodies conducive to them. Fracastoro further specified that some particles spread only through direct contact between bodies, while others could survive for long periods of time without a host.[9] Crucially, however, Fracastoro did not know what originally produced the *seminaria*, though he hypothesized that they might originate in the body or in the outside world as a result of unfavorable planetary alignments.

Borrowing from the tradition established by Fracastoro, the investigators imagined the material cause of yellow fever as a specific, though invisible, particle of matter. As Columbia Professor Richard Bayley wrote in 1796, "By contagion we understand something *peculiar* and *specific*, possessing properties *essentially* different from anything else."[10] The contagionist Cathrall identified three ways through which these "peculiar" and "specific" particles infected hitherto healthy individuals: first, through "immediate contact with the patient's body"; second, through "the matter of contagion arising from the morbid body impregnating the atmosphere of the chamber, and being applied to susceptible constitutions"; and last, from "substances which had imbibed the matter of contagion" and had "the power of retaining and communicating it in an active state, such as woollens, furs, &c."[11] Cathrall likened the contagion of yellow fever to the contagion of the much better understood smallpox, a baneful disease in the eighteenth century. Long experience with smallpox had convinced everyone, even those who most strenuously argued for the local origins of yellow fever, that it operated through contagion.[12]

Unlike contagionism, localism derived from a specific place and time: fifth-century BC Greece and the works imputed to Hippocrates of Cos. Like contagionism, it rested on observations about the patterns of certain diseases. Hippocratic authors acknowledged the existence of a class of diseases, known as fevers, which were not manifestly contagious. Noticing that fevers came tied to local environmental conditions, they conjectured that they spread to humans via environmental corruptions, or *miasmas* (the Greek word for "impurity"), such as the exhalations from marshes. Undoubtedly the fevers to which they referred were forms of malaria (from the Italian for "bad air"), an endemic disease long present in Greece, which is caused by any of a number of protozoa, called plasmodia, and spread through the bites of female mosquitoes of the genus *Anopheles*.[13] Since *Anopheles* mosquitoes prevail in specific environmental conditions, hot and wet areas, the fevers they transmitted appeared to have been caused by those conditions.

In North America localism appealed to the scores of doctors who confronted the deadly ravages of malarial "fevers." Known to Americans as the "bilious," "remitting," "intermitting," or "autumnal" fevers, the *falciparum* and *vivax* varieties of the malaria plasmodia, along with its insect vector, *Anopheles quadrimaculatus*, imposed a harsh reign over the Low Country of the American South. Malaria also struck farther north, in and around cities such as Philadelphia and New York, where it was known as the "autumnal" fever, for its tendency to strike during the hottest and rainiest time of year.[14] Given its environmental specificity, however, everyone agreed that these "fevers" were caused by miasmas. In his health manual for inhabitants of the malarial Low Country, the doctor turned historian David Ramsay claimed that the seasonal fevers arose "from the separate or combined influence of heat, moisture, and marsh miasmata." Even Currie, the most energetic advocate of the contagionism during the yellow fever years, acknowledged the local origins of Philadelphia's "autumnal" fevers.[15]

The fever investigators believed that miasmas possessed physical properties. Countless particles of matter, miasmas emitted from decaying or putrid substances—such as rotting carcasses or the fetid matter in swamps and stagnant waters—and then either hovered in the air, like dust or pollen, or were themselves different types of air. Borrowing the language of Antoine-Laurent Lavoisier's new chemistry, investigators described miasmas as the products of fermentation, the process by which the chemical bonds that held living bodies together dissolved, only to form new, sometimes deadly, combinations. Given their identification with the aeriform state, disease inquirers likened miasmas to odors (they often described them as "putrid" or "noxious" "effluvia" or "exhalations"), which they believed arose from the constitutive elements of the matter from which they came (thus, the characteristic smell of, say, wood came from small pieces of floating wood).[16] One anonymous investigator rather fancifully tried to prove the local origins of Philadelphia's fevers by claiming that he knew someone who could smell the putrefactive miasmas choking the city.[17]

Localism and contagionism offered early republicans compelling models for considering the origins of new and unfamiliar diseases such as yellow fever. Both held that small pieces of noxious material entered the fragile, temperamental human body, inducing states of disease. Moreover, both approaches convincingly and unambiguously explained the causes of certain diseases. Clearly, then, the fever investigators did not simply appeal to their favorite theory in order to explain yellow fever. They were still faced with an important question: Given the existence of two long-standing, widely accepted models of disease transmission and causation, how were they to determine which one applied to yellow fever?

Though they would come to disagree over a great many points, the investigators agreed that the problem of yellow fever demanded "facts." The lexicographer Samuel Johnson defined a "fact" as "a thing done; an effect produced," adding that it could also be described as "reality; not supposition; not speculation."[18] Conceived as distinct pieces of knowledge, or as isolated observations or measurements of nature, facts provided the building blocks of deeper understandings. Crucially, facts enjoyed a "rugged independence of any theory"—they would still be facts even if they did not agree with favorite theories, or if there were no theories to explain them—which suited the dominant epistemological orientations of early American knowledge producers.[19] Since anyone could produce facts, and most could evaluate their meanings, facts negated social distinctions and educational levels. Facts freed natural inquiry from the clutches of entrenched authorities, systematizers, and learned, though frequently misguided, theorists. In a society that was deeply suspicious of authority and disdainful of elitism, the facts linked inquirers to a transparent, egalitarian method of knowledge production, in which truth claims were decided ultimately by the "public"—the collective weight of opinion exerted by the community of knowledge-makers and consumers in a vast republic of letters that spanned the Atlantic.[20]

Still, for all of their value, discerning observers also recognized that facts did not descend unmediated from the phenomena they were supposed to describe. To the contrary, people made facts—or they translated experiences into facts—and so the validity of facts depended on the accuracy and reliability with which they were recorded and on the honesty of those who recorded them. Ultimately, then, the fact culture of the early republic—the "democracy of facts," as Andrew Lewis calls it—rested on an unspoken but widely held assumption (one that would crack and disintegrate during the yellow fever years) that the community of disinterested knowledge-seekers would faithfully and diligently convey the facts that came before them.[21]

For the fever investigators, the facts were discrete parcels of faithfully recorded, firsthand information about yellow fever, such as where, when, and whom the disease struck. Facts enabled investigators to follow the pathways of yellow fever—to trace its routes through cities or from victim to victim. In the disease-stricken cities, where panic and rumors swirled, the facts might just be capable of setting the record straight. In one way or the other, everyone courted the facts. Justifying his choice of causal explanation, Cathrall stated that it was the notion "supported by the greatest number of facts."[22] The localist Bayley also emphasized the centrality of facts: "It is the main object of the writer of this pamphlet to lay before the public a few facts on this important subject; and in doing this, he is conscious that the chief merit of these will consist in the diligence with which they are collected, and the fidelity with which they are detailed." Bayley believed that facts would settle the controversy: "The object I have in view is to reconcile those opinions, as far as that can be effected, by bringing into one view all the facts I have been able to collect on the subject."[23]

Besides highlighting their commitments to facts, the investigators emphasized their disdain for theories. Deceptive and dangerous notions, theories imposed assumptions on facts and frequently led inquiring minds astray. They compared unfavorably to experience. As the localist David de Isaac Cohen Nassy of Philadelphia claimed, "What I am to advance, shall be less founded on theory, which often deceives, than on practice, and my clinical observations. Thus I will only say what I have seen, or believe myself to have seen."[24] Currie likewise stressed the superiority of fact over theory: "Instead of attending to systematic arrangement," a byword for speculative philosophical discourse, Currie "contented himself with enumerating symptoms and circumstances as they occurred to him, while engaged in practice." The result, he hoped, would prove beneficial to subsequent readers, who might one day, too, confront the mystery of yellow fever: "If it should ever appear here again . . . physicians may not again be at a loss for a directory, derived from actual experience and observation."[25]

The investigators' eagerness for facts and hostility toward theory underscored their allegiance to a type of natural inquiry that had its roots in the Scientific Revolution. Knowledge of nature, they asserted, required empirical evidence and inductive reasoning. The philosopher was to seek the evidence of the senses and experience, which he (not, with very few exceptions, she) could then amass in sufficient quantity so as to infer general rules of nature.[26] In medicine, the scientific-empirical turn led inquirers back to the ancient world and the Hippocratic corpus. More than any specific ideas or practices, the Hippocratic tradition promoted general ways of thinking about disease and the body. In widely read works such as *Airs, Waters, Places*, and *Epidemics*, Hippocratic authors urged doctors to mind the environmental conditions of the places where they practiced and to take comprehensive notes of the phenomena that accompanied diseases. Through the advocacy of Thomas Sydenham, the famed seventeenth-century English physician, Hippocratic thought enjoyed a renaissance in the eighteenth century. Nicknamed the "English Hippocrates," Sydenham encouraged his contemporaries to emulate the Greek father of medicine by composing accurate case histories and taking diligent notes of the precise circumstances—the facts—that accompanied each occurrence of disease.[27]

By the eighteenth century, neo-Hippocratic thought had gained wide currency among medical practitioners in Europe and the New World. Medical luminaries from Friedrich Hoffmann to the Italians Bernardino Ramazzini and Giovanni Lancisi incorporated neo-Hippocratic perspectives into their celebrated works; so too did William Cullen, mentor to dozens of American students at the University of Edinburgh.[28] Meanwhile, in the wider Atlantic world, doctors such as James Lind, the popularizer of the prophylactic for scurvy, and countless voyagers and settlers adopted the empirical approach of Hippocrates to help them understand the new and often strange diseases—scurvy, malaria, and of course yellow fever—they encountered onboard ships and in unfamiliar tropical environments. During the Seven Years' War, the American Revolution, and the Haitian Revolution, military doctors stationed in the West Indies regularly marshaled Hippocratic perspectives to answer questions about yellow fever, the dread disease of newcomers in the sultry West Indies.[29]

The investigators were enmeshed in the same trans-Atlantic networks of knowledge exchange through which neo-Hippocratic thought flowed.

An expansive book trade linked early republicans with medical literature from Europe and beyond. With their lending libraries and bookstores, cities such as Philadelphia and New York offered easy access to medical books, including the works of neo-Hippocratic authors, who encountered yellow fever in disparate locales.[30] On a more informal level, medical inquirers in the early republic also forged epistolary links with each other and with doctors in Europe and the West Indies. Letters transmitted vital intelligence about diseases that occurred in remote parts of the world, thus enlarging the number of facts that investigators could bring to bear on the question of yellow fever. Letter-writing also cemented personal bonds among gentlemanly natural philosophers separated by space.[31] Letters constituted integral nodes in the discursive communities that linked inquirers in the United States with those in the Atlantic. Contemporaries acknowledged their importance, but they also recognized that for all letters could build, they could destroy. As the epidemic period proceeded, the private nature of letters became an object of intense suspicion, and a cause of the disintegration of the very communities they helped form.

By the latter part of the eighteenth century, Hippocratic ideas had begun to take root in the earliest American medical schools. Modeled after the medical school at the University of Edinburgh, American medical schools and their teachers adopted the foci of their predecessor institution and its illustrious neo-Hippocratic physician, Dr. Cullen.[32] At the University of Pennsylvania, students learned from Rush, known popularly as both the "Hippocrates of America" and the "American Sydenham" for his advocacy of facts and observations in the study of disease.[33] At King's College (later Columbia), Professor Samuel Bard likewise endorsed Hippocratic ideas: "In the Prosecution of your Studies," Bard told his students, "let such Authors as have transmitted to us Observations founded upon Nature, claim your particular Attention. Of these, HIPPOCRATES shines the foremost."[34]

Hippocratic thought also spread outside academy walls, in numerous clubs and voluntary associations. Like the American Philosophical Society, the prototype of the American intellectual club, medical associations linked intellectuals, and promoted rational inquiry and debate. Associations formed integral units in the early republican public sphere. The most prominent of early republican medical associations, the College of Physicians of Philadelphia cofounded by Rush and Currie, specifically endorsed the Hippocratic approach. As its charter explained, "The objects of this College are to advance the science of medicine, and thereby to lessen human misery, by investigating the diseases and remedies which are peculiar to this country; by observing the effects of different seasons, climates and situations upon the human body; by recording the changes which are produced in diseases, by the progress of agriculture arts, population and manners."[35]

The mission statement of the College of Physicians broadcast the high hopes and ambitions of early republican medical inquirers. The fevers struck at a time of soaring confidence, when Americans, fired with faith in reason and progress, fully expected to usher in a new age of science and medicine. Joel Barlow indicated as much in his *Vision of Columbus*, in which he predicted the death of "pale diseases" in America: "And blooming health adorn the locks of snow, / A countless train the healing science aid, / Its power establish and its blessings spread."[36] Rush similarly foresaw a new era dawning over the United States. "All the doors and windows of the temple of nature have been thrown open by the convulsions of the late American revolution" and "human misery of every kind is evidently on the decline."[37]

These sanguine predictions rested on equally positive estimations of the healthfulness and vitality of American environments. Bristling at the Comte de Buffon, who criticized the degeneracy of American environments, animals, and people, American naturalists such as William Barton demurred. Owing partially to "salubrity of the climate," Barton wrote, "this country possesses, in a superior degree, an inherent, radical and lasting source of national vigor and greatness."[38] Currie also celebrated the United States for its lack of disease: "The diseases which do occur are more simple and uniform; and this country is intirely exempt from some of the most formidable and destructive which infest the other quarters of the globe." He specifically praised Philadelphia: "The chances of enjoying health, and prolonging life, is much greater in the City of Philadelphia, and some other parts of the united States, than in any other districts of the world."[39] In the minds of these early republicans, the American *novus ordo seclorum* would not only perfect politics and civil society, it would also transform health.

Drawn to the facts for their basis in empirical and inductive epistemology, and for their tendency to promote transparent, egalitarian knowledge production, the investigators soon learned that facts opened themselves to many interpretations. Thus, while they might agree on what the facts were, they disagreed markedly over what they meant. For instance, everyone agreed, it was an undeniable *fact* that yellow fever conformed closely to the boundaries of the affected cities, and seldom strayed far beyond. "It was a singular fact, that when carried in the country, it never was, but in one instance, that I am acquainted with, propagated beyond the person who carried it," Cathrall acknowledged.[40] The localists took the spatial limitations of the fever as decisive proof that it was spread not through contagion but by means of a miasmatic vapor situated locally within the cities. Writing to John Morgan of the College of Physicians, Rush averred that the disease's failure to move outside of Philadelphia in 1793 obviously meant that it had been "deprived of the aid of miasmata from the putrid matter which first produced it in our city."[41]

Localists drew similar conclusions from the time of the season and the climatic conditions during which yellow fever struck, another undisputable fact about the disease. Each year, the epidemics began in the late summer and continued through the autumn, the period of the year when heat, humidity, and heavy rains—the acknowledged ingredients of pestilential fevers—afforded putrid matter opportunities to develop into deleterious miasmas. Quoting a letter from Dr. Edward Miller, future cofounder of the *Medical Repository*, Rush wrote that the epidemic in 1793 occurred when the climate nearly resembled a "TROPICAL season" (1793 was an El Niño year!). In consequence, Rush continued, "We ought not to be surprised if tropical diseases, even of the most malignant nature, are ENGENDERED amongst us."[42] Each year it struck, yellow fever prevailed at the exact time when the common remitting, or autumnal, fevers usually struck. Since the remitting fevers were known to be the effects of the environment, localists conjectured that yellow fever must only represent a higher degree of the common fevers, and that it must, therefore, also derive from the surrounding conditions. The symptoms of each of the fevers did bear certain striking resemblances—high fever, lassitude, nausea, and stages marked by remissions.[43]

The contagionists looked at the very same facts, but reached different conclusions. Yellow fever could not be caused by environmental conditions, they argued, because it induced symptoms unlike the "native" diseases of the West Indies and American South. Joseph Mackrill, a physician from Baltimore who had practiced in the West Indies, conceded that though yellow fever began like the native diseases of the West Indies, patients quickly manifested unusual symptoms, especially, he noted, a "slow and oppressed" pulse.[44] In his *Observations on the Causes and Cure of Remitting or Bilious*

Fevers, his second work on the malarial fevers, Currie further highlighted the singular features of yellow fever. "The malignant yellow fever is distinguished at its commencement, from the worst cases of the bilious remitting fever, by the suddenness of its attack . . . greater severity of pain in the forepart of the head and eyes . . . and especially by the costiveness or dysenteric state of the bowels." More to the point, yellow fever terminated in truly unique fashion—"the bilious colour of the skin, and the coffee ground, or black vomitings." Far from being a relative of the bilious remitting fevers of tropical or semitropical climates, yellow fever's distinctive symptoms made it unlike any other disease, except, Currie concluded ominously, the "plague."[45]

Localists and contagionists alike also noticed (it was a *fact*) that former inhabitants of the West Indies largely escaped the disease, while those from temperate climates fell victim with astonishing regularity. Most also noted that Africans possessed some special ability to resist yellow fever. At the outset of the 1793 epidemic in Philadelphia, Rush even encouraged members of the black community to volunteer for service at the makeshift hospital, believing that they were naturally immune. He later realized his mistake, but he still maintained that blacks did not acquire yellow fever as frequently or as severely as whites. Cathrall likewise remarked that "blacks of every description, were less liable to it than the white inhabitants; and the negroes originally from the coast of Africa were scarcely ever affected." In New York, Valentine Seaman claimed that blacks acquired yellow fever as frequently as whites, but that it was not "so fatal to them."[46] Good evidence suggests that they probably were correct. As inhabitants of the continent where yellow fever originated more than 3,000 years ago, Africans from yellow fever endemic zones probably had acquired some degree of genetic immunity to the disease.[47]

Some interpreted the selectivity of yellow fever as a certain indication of its local origins. As Hippocratic thinkers, localists liked to point out that the human body existed in a constantly evolving, dynamic manner with its surroundings. If West Indians and Africans resisted yellow fever better than the strangers to tropical climates, it was only because their bodies had adapted to the environmental conditions that produced the fatal malady. On the other hand, as pieces of noxious matter totally unrelated to the environment in which they occurred, contagions theoretically should have attacked all human beings equally. According to the localists, observation had confirmed this suspicion. "It is an acknowledged fact, with respect to contagions in general," Seaman declared in his treatise, "that they are no respecters of persons, but that all of every clime and color, under like circumstances, are equally susceptible of their operations."[48] Yellow fever, it would seem, could not be contagious.

The contagionists simply altered the direction of the argument. Commenting on the "notorious fact" that West Indians and Africans escaped the worst effects of yellow fever, Currie claimed it only proved that the disease differed fundamentally from the remitting fevers, which, everyone knew, affected all people equally. Currie thus neatly undermined a major pillar of the localist view, the notion that yellow fever was only a higher grade of the common fevers (i.e., malaria). Since yellow fever was different from the malarial fevers, then it must be its own unique disease, native to the West Indies, and therefore imported to the United States.[49]

Investigators offered opposing interpretations of the facts, but the contagionists did accept some elements of the localist interpretation. At least at the beginning of the epidemic period, Currie for example was willing to concede that local conditions might activate, or exacerbate, the contagious particles, just as a seed requires nourishment to germinate. "Though it was propagated by contagion," Currie conceded in a 1794 treatise, "the sensible qualities of the atmosphere had a surprising effect in rendering the contagion more or less active."[50] The admission seemed reasonable given the facts—there must have been something about the atmosphere of cities that made contagions more robust—but tellingly Currie backed away from such claims as time wore on and opinions hardened around causal models. The localists never offered such compromises.

Whatever their initial concessions, the contagionists' arguments centered on one overarching and undeniable fact: they could always trace the first appearances of yellow fever to incoming ships that carried passengers with yellow fever. Writing of the 1793 epidemic, Mathew Carey correlated the first cases of the disease in early August with the arrival of three ships in late July—the *Amelia*, from Saint-Domingue, and the privateer *Sans Culotte* with its prize, the *Flora*. Carey even demonstrated that the *Amelia* docked at the wharf near the house of the first reported victim of the epidemic, a woman known as Mrs. Lemaigre.[51] Subsequent investigators did not have to look hard for similar concurrences of events. In 1795, the health officer of the port of New York, Malachai Treat, visited a vessel called the *Zephyr* from Port-au-Prince, which he ordered into quarantine after noticing that it carried victims of yellow fever. When Treat sickened and died with the same fever that was then erupting in the city, numerous observers simply concluded that the *Zephyr* had imported yellow fever, that Treat contracted the disease, and that it then spread through the rest of the city.[52] In Philadelphia in 1798, members of the College of Physicians similarly linked the arrival of the *Deborah*, an infected vessel from Saint-Domingue, to the following epidemic.[53] For contagionists, the coincidence of diseased ships and yellow fever outbreaks seemed clear evidence of cause and effect.

So much for the facts. The facts reduced the debate over yellow fever into a stalemate, with both sides having marshaled compelling arguments supported by strong evidence. The facts themselves only reflected the confounding etiology of yellow fever. Striking according to the vagaries of its insect vector, the pesky *Aedes aegypti* mosquito, yellow fever behaved strangely. The contagionists were right to point to the arrival of infected vessels—yellow fever did require importation, because *A. aegypti* could not survive the winters of the northeastern seaboard. But the localists were correct to implicate the environmental specificity of yellow fever and the conditions of the port cities—high temperatures, rainfall, and the abundance of water receptacles did ensure the survival and propagation of the mosquitoes. But the question was: If not with the facts, then how could they determine what caused yellow fever?

Had they been as stringent about the facts as they advertised, investigators might have acknowledged that no one interpretation of the facts was any more convincing than the other. As it was, they did not all depend on the facts as much as they claimed or wanted their audiences to believe. It was not that facts were flawed, nor that the investigators feigned their commitments to empirical evidence and inductive reasoning, the hallmarks of their scientific heritage. Rather, leading intellectuals such as Rush, Mitchill, and Webster balked at the perceived excesses of scientific thought, especially its association with "empiricism." A theory of mind with sinister implications, empiricism rejected the innate rational capacity of the human mind, a gift of God, according to those who believed in it, and a scientific truth in its own right. In order to redeem scientific inquiry, the localists— Rush, Mitchill, Webster, and others educated on the Scottish model— championed the union of science with common-sense philosophy, and they imposed this construct over the yellow fever debate. With their humbler educational backgrounds, however, contagionists never evinced any particular allegiance to common sense. More than any other factor, common sense differentiated localists from contagionists.

Western thinkers had long believed that the human mind possessed innate powers of reasoning. Beginning in the late seventeenth century, philosophers began to reject the time-worn mental theories in favor of a new picture of the mind, and its severely limited capacities, known as empiricism. In this particular form, empiricism represented a species of thought first articulated by John Locke in his *Essay Concerning Human Understanding* (1690) and then adopted by later British philosophers such as George Berkeley and David Hume, and, in the French tradition, by Étienne Bonnot de Condillac. The "empiricists" held that humans were born without innate ideas. Locke used the memorable and oft-repeated phrase *tabula rasa*, or "blank slate," to describe the condition of the mind at birth. The empiricists believed that humans accumulated all of their ideas, all of their knowledge, through sensory impressions, which the mind then stored away and categorized. Hume, the most skeptical of the empiricists—and the vilest in the minds of later detractors—went so far as to argue that empiricism negated the human ability to understand cause and effect. Since the mind grasped nothing a priori, and since all it contained came from experience, Hume reasoned, then everything that humans thought they knew about causation must really only be the product of events "constantly conjoined" in time and space. "We only learn by experience the frequent CONJUNCTION of objects, without being ever able to comprehend any thing like CONNEXION between them."[54] In other words, when one event or "object" is always followed by another, we assume that they have a causal relationship, without truly understanding the force or power that produces it.

Hume's skepticism, embodied in his theory of the mind, sparked a cross-denominational backlash by Christian natural philosophers in both Europe and North America. The empirical view of the mind threatened long-held and cherished notions about the nature of God and creation. To suggest that the mind was an empty vessel that could only store away empirical data meant that the human body—that "minute fabric, a center of wonders," as Rush described it—was a mere machine that possessed no remnant of its divine origin. What is more, by denying innate ideas, Hume also the rejected the human ability to comprehend God's design, the ultimate end of science for the devout. Hume's empiricism left humans as blank automatons in a world that they could never truly understand. More ominously, if taken to its extreme, it amounted to a rejection of God's benevolence. Rush grasped its irreligious implications: "It would be a denial of goodness to the Supreme Being to suppose, he had not endued the common faculties of man, with the means of discovering, and obviating the common physical evils of his life."[55] In his own *Medico-chymical Dissertations on the Causes of the Epidemic Called Yellow Fever*, Felix Ouvière likewise remarked, "The capacity to acquire a sufficient knowledge of the laws of nature, is a gift bestowed upon us by the Supreme Ruler, to the end that we may derive from it all those blessings which it is susceptible of yielding."[56]

The alternative came with the common-sense philosophy of the Scots. The Scots did not invent common sense. It had existed since the beginning of Western philosophy—Aristotle referred to a "common sense" that existed among humans—and in the eighteenth century alone, common sense figured in discourses about innate human sociability and morality, and people's abilities to make reliable political judgments (this was how Thomas Paine used the term in his pamphlet of the same name).[57] But it was the Scottish philosophers, especially Thomas Reid of the University of Glasgow and James Beattie of Aberdeen, who molded common sense into an epistemology in response to the heretical skepticism of Hume. Reid and Beattie argued that the mind did not merely accumulate sense impressions but also possessed innate powers of reasoning. Reid's common-sense philosophy, for example, held that "some original principles of our constitution" made "reason and experience possible."[58] To him, the very idea that humans could attempt to use reason and gather experience suggested that the mind possessed inherent powers. As an epistemology, common sense broadened the standards of certainty. According to James Delbourgo, common sense reaffirmed "the capacity of ordinary human beings to make reliable judgments about natural phenomena."[59] It suggested that humans could verify truths of nature independently of strict empirical evidence.

Common-sense philosophy flourished in the American colonies and the United States, long a bastion of Scottish learning. Most Americans found that common sense jibed with religion. Since common sense held that certain truths simply exposed themselves to the common perceptions of man, it buttressed evangelical religious devotion—the dominant and unifying strain of American Protestantism in the early republic—which stressed experiencing or feeling one's spiritual connection with God. The corresponding triumph of "common-sense moral reasoning" encouraged Americans to build their ethical beliefs and behaviors through independent reflections on their innate moral senses, rather than through the traditional avenues of religious authority.[60] Common-sense philosophy proved equally amenable to scores of devout American natural philosophers who used it to escape the crushing doubt of Hume and, they thought, the cloudy sophistry that only hindered the pursuit of useful knowledge. Common sense also vindicated the goodness of God, because it showed that the creator had given his creation the ability to combat the evils of the world, such as yellow fever. From the revolutionary period on, common-sense philosophy increasingly infiltrated America's Protestant academies, its universities, and its intellectual circles.[61]

As leading intellectual figures, many of the fever investigators ranked among the most vigorous proponents and disseminators of common-sense philosophy. "Reverberate over and over my love to Dr. Beattie," Rush wrote to a former student traveling to Aberdeen. "I cannot think of him without fancying that I see Mr. Hume prostrate at his feet. He was the David who slew that giant of infidelity."[62] In his *Treatise on the Autumnal Endemial Epidemick of Tropical Climates, Vulgarly Called the Yellow Fever*, John Beale Davidge of Baltimore, a physician trained at the University of Glasgow (and the future founder of the medical school at the University of Maryland), likewise signaled his debt to apostles of common sense. Comparing contagionists to the followers of false philosophies—the physiognomy of Johann Lavater and, of course, the empiricism of Hume—Davidge, also an avowed Episcopalian, averred that his own ideas fell in line with common sense. "Were I to plunge into the vacuum of metaphysicks," Davidge wrote, "I should believe, with the peerless Reid of Glasgow, that the human mind possessed, inherently, action, vigour, and choice; that it operated upon surrounding objects, and was not the passive sport of incidental impression."[63]

Rush celebrated common sense for its ability to guide philosophers through the labyrinths of knowledge production and expose truth (Figure 1.2). In an essay called "Thoughts on Common Sense," Rush articulated his own version of common-sense philosophy. Referring to common sense as "the perception of things as they appear to the *greatest* part of mankind," he nevertheless asserted that humans possessed an inborn rational capacity, which he identified as "reason," and that it exerted a powerful influence over

Figure 1.2. Dr. Benjamin Rush by Charles Willson Peale, 1783–1786. In this portrait, Rush is composing a lecture on the causes of earthquakes. Barely visible in the background on his bookshelf are a volume by Hippocrates and "Beattie on Truth," a reference to Beattie's common-sense opus, *An Essay on the Nature and Immortality of Truth, in Opposition to Sophistry and Scepticism* (Edinburgh, 1770).
SOURCE: Courtesy, Winterthur Museum, gift of Mrs. Julia B. Henry. 1959.160.

the way that they gained knowledge of nature. "The principal business of reason," Rush wrote, "is to correct the evidence of our *senses*." He went on, "The perceptions of truth . . . consist in little else than in the refutation of the ideas acquired from the testimony of our senses." Rush's elevation of "reason" subordinated empirical evidence, the facts, to the organizing power of the mind. Left to its own capacities, the mind could independently sift through evidence and produce the "perceptions of truth," almost as though these "perceptions" were sensations themselves that would indicate to the thinker when he had discovered "truth."[64]

Above all, investigators sought the proper harmony of facts and common sense, empiricism and rationalism. In a short pamphlet discussing the faculty and curriculum at Columbia College, Samuel Latham Mitchill proudly noted that his own institution and its illustrious professors—Samuel Bard, Wright Post, William Hamersley, and Richard Bayley—had struck just the right balance between empiricism and rationalism. Borrowing a long passage from Francis Bacon's *Cogitata et Visa* (1607), a key text in the rise of scientific epistemology, Mitchill detailed the college's "middle way" with a metaphor from the animal kingdom:

> Faculties of the arts and sciences, whether empiricists or rationalists—all philosophers must agree—Empiricists in the manner of ants collect in order to put to use, while rationalists in the manner of spiders spin webs from themselves. The bee is the middle way—it takes materials from the flowers of the gardens and the fields, but it manipulates and distributes them with its faculties. Not unlike the work of real philosophy, which from natural history and experiments, takes material and puts it away, not in the memory as a whole, but in the mind in a developed and changed way.[65]

Rather than simply storing away ideas and information, the fertile human mind, like an industrious bee, gathered them together and adapted them to its purposes. The metaphor allowed Mitchill to describe the proper balance between empirical evidence and common sense, both of which were necessary for the construction of true knowledge. It also came with the implicit approval of Bacon, the widely recognized creator of scientific philosophy, thus placing Mitchill's mixture of empiricism and rationalism in a lineage that went all the way back to the beginnings of the Scientific Revolution.

From the very beginning, proponents of common-sense philosophy tried to establish it as a type of science in its own right. Reid, Beattie, and their American cohorts fully embraced the empirical and inductive methods of the latest science; they only differed from the arch-empiricists in terms of their theories of the mind. Reid even attempted to prove the existence of innate ideas inductively, by comparing the similar syntactical and grammatical structures of different languages.[66] The common-sense philosophers fancied themselves natural philosophers of the mind, overturning the baseless metaphysical ravings of "sophists," such as Hume. "The works of Dr. Reid and Dr. Beattie have produced a revolution in the science of metaphysicks in our American seminaries," Rush noted with satisfaction. "It is now very properly limited to the history of the faculties and operations of the human mind." Rush, in fact, contributed his own "Lectures on the Mind" to this emerging field of *scientific* study.[67] Rather than signaling a return to the philosophies of thought alone, common sense moved forward, transforming guesses and speculations into a sort of proto-psychology. Slowly, and rather slyly, the adherents to common-sense philosophy began to turn the tables on Hume and his severe epistemology. By trumpeting their own scientific credentials and denigrating the empiricism of the mind as backwards, medieval metaphysics, Americans sought to legitimize a rival type of knowledge-making.

In the context of the yellow fever debate, common-sense philosophy licensed investigators to exercise their reasons, and it encouraged them to trust their inclinations and intuitions. It inspired investigators with the faith that they could and would convince others of the truth, even if the facts themselves could not. All human beings possessed common sense, after all; the investigators had only to appeal to their reason and they *would* recognize the truth. Rush, for his part, expressed utmost confidence in the eventual success of the truth-seekers—he might have written "localists"—for, he wrote, "They move by the light of reason, the advantages of which in medicine, compared with solitary and mechanical experience, are like the extensive benefits the science of navigation has derived from the loadstone, compared with the feeble aids it formerly derived from the sight of land, or the transient light of the stars."[68] The loadstone (or "lodestone"), the magnetic mineral used in compasses, did make for a pertinent analogy. Like a compass, reason (i.e., common sense) could offer guidance when all was dark and the way forward uncertain. Compared to "solitary and mechanical experience," a reference to experimental science, with its strict empirical edge, common sense provided a crucial supplement, for it pointed out the proper direction and it did so innately.

Common sense established a framework for the consideration of different interpretations of the same evidence, and thus it helped investigators locate dire philosophical flaws in their opponents' perspectives. It suggested that the truth of one's argument about the cause of yellow fever rested not on the facts alone but on the argument's superior plausibility, especially its ability to fit as a comprehensible element in the God-constructed world. If the investigators could trust their senses and inclinations about the natural world—if, indeed, God constructed nature so that humans could understand it—then the world must ultimately be a comprehensible place, and the things in it sensible and consistent. And yet it seemed so obvious (or perhaps "common sensical") to localists that contagionists violated essential philosophical rules about the consistency and coherence of the natural world. Crucially, the localists' critique centered on their opponents' apparent disregard for the laws of cause and effect—oversights which, if accepted, would authorize a type of unphilosophical error just as egregious, impious, and unacceptable as Hume's earlier denial of cause and effect.

So, besides arguments from facts of each occurrence, localists took issue with the contagionists' chief factual argument—the arrival of disease-laden vessels. Localists acknowledged the arrival of disease-laden vessels before outbreaks, but they denied that the ships had causal relationships with the subsequent epidemics. The contagionists, they countered, had fallen into a popular mistake. The arrival of infected ships and the outbreak of pestilence were, as Hume would have put it, merely events conjoined in space and time, utterly lacking in "connexion." Rush used the contagionists' fallacy to lecture about proper philosophy. *"Accidental Coincidence,"* he wrote, "is a frequent source of error." As an example, Rush noted, "A pestilential fever which accidentally succeeded the introduction of the potatoe into France, produced an edict against the cultivation and use of that wholesome root by the French court." Rush then groped somewhat awkwardly to the expected conclusion, "In like manner, the arrival of a ship from the West Indies, and the sickness or death of a sailor induced by the putrid exhalation of our docks and wharves, occurring in the months of July or August . . . has unfortunately [been] connected . . . together as cause and effect."[69] Rush asked his readers to disregard an undeniable, and indeed provocative, coincidence as a simple conjunction of events, mistakenly interpreted as cause and effect.

The localists detected another flaw far more dire in the contagionist argument, one whose full implications materialized only later in the epidemic period. If yellow fever in the United States were caused by an imported contagion from the West Indies, and if, presumably, yellow fever in the West Indies were caused by the same contagion imported from some other place, then where and when did the contagion originate? "It requires no uncommon depth of logic to prove," Charles Caldwell said to an audience of his peers at the Academy of Medicine of Philadelphia in 1798, "that pestilential diseases, having an existence, must have also a place of origin."[70] Yet, the contagionists could not find one, or perhaps more accurately, they never bothered to conjecture about it. Taken to its logical limits, contagionism degenerated into an infinite regress in which disease had no cause but cycled continuously around the world, striking its victims again and again.

Contagionists did not overtly embrace common-sense philosophy, but that did not stop them from detecting philosophical flaws in the localists' arguments. They consistently critiqued localists for indulging in theoretical speculations. Currie attributed localist notions to "Imagination, and her whimsical daughter, *Theory*."[71] Contagionists especially decried the causal inconsistency implicit in the localist argument. If yellow fever sprang from the environment, then why did it not occur wherever those environmental conditions were present, and why did it not remain in some level of endemicity?[72] Natural laws were supposed to operate uniformly, and causes were supposed to produce predictable effects. Isaac Briggs, a correspondent to the *Medical Repository*, perfectly summarized the point, "If a something, proceeding from the putrefaction of animal and vegetable substances, be the parent of malignant fever, and if the same cause uniformly produce the same effect, why are our cities *sometimes* desolated by yellow fever, and not always so, when they contain masses of putrefying substances equally great and numerous?"[73]

The investigators' first inquiries into the cause of yellow fever devolved into controversy. The facts supplied investigators with crucial bits of scientific evidence that combined together in proper inductive fashion resulted in two coherent pictures of yellow fever's origins. Eager to point out the common-sense roots of localism and pry scientific inquiry from the hands of skeptics, localists embarked on an ambitious campaign to discredit contagionism and establish the supremacy of a kind of common-sense science. They alleged that contagionists posited an unwarranted, unsubstantiated causal connection that reduced science to an epistemological morass—with mere conjunctions of events being passed off as scientific truths, and with actual causal relations being pushed further and further back into infinite regresses—exactly as Hume had depicted it. The contagionists did not pursue so ambitious a philosophical project, but we still might detect an anti-Humean element in their reproaches of the localists. If certain features of the environment were capable of producing yellow fever some of the time, then the principles of causality mandated that they *must* do so always, lest the investigators fall into the epistemological void about which Hume had warned. Partisans on both sides of the debate emphasized flaws in their opponents' perspectives that threatened much more than the truth about the cause of yellow fever—they also imperiled the integrity of science and the human ability to understand nature, as the pious early republicans imagined it. As the epidemic period proceeded, investigators found themselves enmeshed in a vexing riddle whose solution held consequences for the health of their cities, as well as the sanctity of their science.

"Declare the Past"

> These records of wars, intrigues, factions, and revolu-
> tions, are so many collections of experiments, by which
> the politician or moral philosopher fixes the principles
> of his science, in the same manner as the physician
> or natural philosopher becomes acquainted with the
> nature of plants, minerals, and other external objects,
> by the experiments which he forms concerning them.[1]
>
> DAVID HUME, *1748*

In 1799, Noah Webster produced a curious and largely forgotten book called the *Brief History of Epidemic and Pestilential Diseases*. Drawn from an array of primary and secondary sources, dutifully examined in libraries ranging over the entire Eastern Seaboard, the *Brief History* (which, at two volumes and more than seven hundred pages, was anything but brief) offered no less than a compendium of every significant epidemic that, to his knowledge, ever struck the world. Webster devoted particular attention to the better-documented epidemics in Western history, such as the biblical plagues; the Athenian plague of the fifth century BC, famously memorialized by the great Thucydides; the pestilences that assailed the Roman Empire during its decline in the late-second and third centuries AD, and during the reign of Justinian in the sixth century AD; the Black Death; and many others. The book also contained comprehensive descriptions of the major climatic, environmental, and astronomical events—earthquakes, volcanic eruptions, and eclipses—that, in Webster's mind, appeared to accompany great pestilences.[2]

The *Brief History* marked a significant antiquarian achievement, but it was no mere bookish enthusiasm that propelled Webster into the study of past diseases and the phenomena that accompanied them. Barely in his forties, and still years from his more famous work as a lexicographer, the patriotic New Englander and federalist firebrand was drawn to the study of disease by the recent spate of yellow fever epidemics. An attempt to put an end to the fractious debate, the *Brief History* finalized Webster's effort to ascertain the true cause of yellow fever. And Webster was not the only one to scour history for answers. As yellow fever ravaged the port cities, and the controversy between localists and contagionists deepened, investigators increasingly turned to the annals of disease history to find the cause of the disease. It started with a few references to historic epidemics, scattered in the investigators' treatises on the cause of yellow fever, and progressed into lengthier, more exhaustive historical studies. By the end of 1799, the investigators' historical research resulted in two remarkable books on the history of disease published nearly simultaneously: the *Brief History* by the localist Webster, and the *Treatise on the Plague and Yellow Fever*, an equally lengthy and exhaustive world history of disease written by an unknown newcomer to the United States, the contagionist James Tytler.

Works such as Webster's *Brief History* and Tytler's *Treatise* testify to the well-noted prominence of history both as a formal genre and a multifaceted element in the intellectual life of the early republic. As the young republic set out on its own, history served as both a source of national identity and guidance. While American intellectuals certainly discovered much to be admired and imitated in history, especially from the classical world, they also found much to be avoided. Political thinkers especially hoped that knowledge of history would enable them to avoid the fates of past republics. Thinking about republics forced them to take heed of history's circularity, its unexpected turns of fortune, its cycles of prosperity and decline, growth and decay, and even health and sickness. In their attempts to escape the past, republican theorists never totally dispelled the fear that such an escape was impossible. But Americans, historians chief among them, still reflected positively on what was different about their situations. Early republican histories tied Americans to an imagined past that stressed the exceptionalism of America. As popular historians such as David Ramsay and Mercy Otis Warren cast it, America's expansive lands, its autonomy from Old World institutions, the plain virtue of its citizens, and the magnificence of the

Constitution all made the United States an exception to the rules that governed the rise and fall of polities, and these qualities promised to sustain the republic through the vicissitudes of time.[3]

The popularity of disease histories also showcases the little-known importance attached to history as a tool in the scientific study of disease. A vast storehouse or repository of facts and observations, history appealed to the investigators' desire to place the study of nature on a firm empirical footing. It suggested that if the habits of diseases like yellow fever could be catalogued and compared, then through a kind of inductive process, investigators could ascertain the cause of yellow fever. In the classic stories of disease, such as Thucydides's plague narrative, the investigators found the very data that would put the study of disease on a sound scientific foundation. If history repeated itself, the investigators reasoned, then by studying it, by using it, they could free Americans from the cycles that doomed peoples and nations to chaos and disease, and avoid the fates that had befallen the less enlightened. Just as early republicans used history to escape the past and create their *novus ordo seclorum*, investigators used history to escape from the cycles that doomed people to disease and to avoid the fates that had befallen the less enlightened. The appeal of history in early republican thought was even broader than scholars have acknowledged.[4]

In the contest between Webster and Tytler, localists tipped the balance in their favor. Webster concluded that while yellow fever certainly took its power from local emanations, its occurrences depended ultimately on a more sinister force—an elusive, occult quality to the atmosphere called the "epidemic constitution," the same mysterious force that produced the other noteworthy periods of epidemic disease, as well as the strange natural phenomena that accompanied them. His explanation made the local production of diseases appear like a normal and plausible element of the natural world, and it placed the United States coherently into the sweep of history. Meanwhile, Tytler's work met with censure from his philosophical contemporaries, who chided him for his assertion that disease originated from divine intervention.

The victory for the localists, however, never went uncontested, nor did it come without a price. Even as the fever investigators combed the records of the past, they uncovered gaping holes in the evidence, inaccuracies, contradictions, and outright lies—weaknesses that undermined history's ability to serve as the empirical foundation for the scientific study of diseases.

Furthermore, though localists might celebrate history's ability to explain the miasmatic origins of diseases and the epidemic constitution, even Webster had to admit that he had not pinpointed the precise causal relationships that led to outbreaks of yellow fever. He had only located a rough correlation among natural phenomena and disease, which appealed to common sense but did not prove that they shared a causal relationship. Worse still, if there was an epidemic constitution—a fundamentally inscrutable force of nature that could rend the earth, incite volcanic eruptions, and produce epidemic disease—what did that suggest about early republicans' conceptions of themselves and their supposedly exceptional place in history?

From the very first appearance of yellow fever in Philadelphia in 1793, early republican intellectuals instinctively compared their calamity to those of the past. Writing to his wife on August 25, 1793, only days after the first appearance of yellow fever, Benjamin Rush noted, "This morning I witnessed a scene . . . which reminded me of the histories I had read of the plague."[5] Mathew Carey similarly likened the public reaction to the epidemic to the best-known epidemic in recent European history, the plague of London in 1665–1666. "It is not probable," Carey wrote in his *Short Account*, "that London at the last stage of the plague, exhibited stronger marks of terror, than were to be seen in Philadelphia, from the 25th or 26th of August, till pretty late in September." Likewise, in emphasizing the extreme mortality of the epidemic, Carey claimed, "The plague of London was, according to rumour, hardly more fatal than our yellow fever." In the fourth and final edition of the *Short Account*, he even appended short descriptions of the plagues in London and in Marseilles in 1720.[6]

Carey's allusions to the plague of London actually reflected a much broader public interest in historic epidemics. In the fever-stricken city and beyond, newspapers teemed with references to past epidemics. One anonymous contributor to the *Federal Gazette* recommended burning tobacco as a preventative against the fever, noting that "when the great plague raged in London, in 1665, it was found that the street where the tobacconists lived was exempt from the general calamity."[7] Hawkers of medicine likewise used history to bolster the effectiveness of their yellow fever nostrums. In a regular advertisement, the shopkeepers Goldthwait and Baldwin and the "druggist and chemist" John White touted the historical lineage of a cure known as the "Vinegar of the Four Thieves." "It is said," they claimed in their advertisement, "that during the dreadful plague at Marseilles, four persons, by the use of this essence or salt, as a preservative, attended, unhurt, multitudes of those who were infected; that under the colour of those services, they robbed the sick and the dead; and that one of them, being afterwards apprehended, saved himself from the gallows by discovering the secret:—The preparation was hence called, VINAIGRE DES QUATRE VOLEURS—or, the VINEGAR OF THE FOUR THIEVES."[8]

From the start, the fever investigators also used the history of diseases in their treatises, though principally as a tool in the search for the causes of diseases. The localists seized upon history as a means of understanding the circumstances under which past epidemics occurred. Localists, recall, believed that yellow fever belonged to a class of diseases, called "fevers," which were caused by the putrid miasmas that emanated from decaying substances. If they could show that other types of "fevers" arose from similar environmental circumstances, particularly decaying matter, then they could argue that yellow fever, by analogy, must also arise from those circumstances. "If we carefully peruse the history of diseases," Richard Bayley, future health officer to the port of New York City, wrote, "we shall find, that those which have proved most fatal to the human race, have proceeded, either directly or indirectly, from some acknowledged peculiarities of the state of the air."[9]

Drawing from his own careful historical research, Benjamin Rush concurred. In his *Enquiry into the Origin of the Late Epidemic Fever in Philadelphia*, published in December of 1793, Rush compared the rise of yellow fever from putrefying coffee to the rise of a fever in Rome from rotting hemp, as described by Giovanni Lancisi, the great Italian doctor and an early apostle of Hippocratic environmentalism. Rush searched history for other examples of fevers emitting from the putrid fermentation of vegetable matter and producing symptoms similar to those of yellow fever. An epidemic in Switzerland, described by Albrecht von Haller in his *Physiological Elements of the Human Body* (1757–1760), occurred under similar conditions and produced similar symptoms to those of the recent yellow fever outbreak in Philadelphia. Another "malignant fever" at Oxford College was exhaled from a "vast quantity of Cabbages," killing many of the students; and putrefying "radishes, turnips, garlic and sundry other vegetables" produced similar effects in other instances.[10] For Rush, the similarities between yellow fever and the diseases in question seemed sure evidence of their common origin from the effluvia of decaying matter.

With all of their variations in terminology, scope, and emphasis, the descriptions of past diseases could very easily be misinterpreted, though often in constructive ways. When viewed from a certain perspective, most any disease could appear to be yellow fever. In his dissertation at the University of Pennsylvania in 1804, Stubbins Ffirth maintained that James Cartier described yellow fever among the Indians in Canada at the time of French arrival, and that John Winthrop wrote about the same disease among another group of Indians in 1635. Having himself examined the firsthand testimony of Native Americans from Dan Gookin's *Historical Collections of the Indians in New England* (1792), Ffirth also discovered evidence of yellow fever afflicting the "Pawkunnawhutt" Indians, who claimed to have exhibited generalized yellowness as a symptom of one of their diseases.[11] Since yellow fever seemingly preceded European contact with the North America, he concluded that it originated from local sources, not foreign importation.

Contagionists, too, buttressed their arguments with the evidence of history. Like the localists, contagionists constructed their earliest historical arguments from analogy. If they could show that past diseases that resembled yellow fever (perhaps even yellow fever itself) arose from contagion, they could make an argument in favor of its contagiousness in the United States. John Beale Bordley claimed that the yellow fever epidemic in Baltimore in 1794 resembled the "*Natolian* plague" of Constantinople, Egypt, and the Barbary Coast, which, according to the travel writer M. Savary, usually arose from importation. Since the plague, a close relative of yellow fever, originated from importation, so too must yellow fever. Struck by the value of the past, Bordley admonished his colleagues that they would learn much from the "histories of the disease heretofore published."[12] William Currie, likewise, asserted that the "plague of Athens, described by Thucidides," was "exactly the same distemper" as the yellow fever of the United States. Since the Athenians themselves believed their plague to have been imported from Ethiopia, Currie predictably concluded that the Americans had also imported their disease.[13]

The historical lens clearly disguised the differences among diseases. Taken in isolated examples, the evidence gleaned from historical disease outbreaks would never turn the tide in favor of one causal explanation or another. Nor could appeals to history answer the philosophical problems left by the earliest treatises. History, an undeniably intriguing repository of information, was also a shifting one, as the investigators were beginning to learn. In order to produce persuasive results, historians of disease would have to look deeper at the past. Perhaps then contagionists could find some end to the regress implied by their doctrine, and the localists some rationale for the apparent causal inconsistency implicit in theirs.

The turning point came with the epidemic that struck New York in 1795 (Figure 2.1). "The disorder first made its appearance in July," Matthew Livingstone Davis wrote in his *Brief Account of the Epidemical Fever*, a small pamphlet published in the immediate aftermath of the outbreak. By the

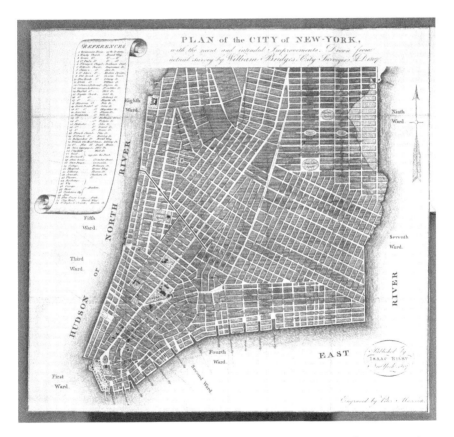

Figure 2.1. Plan of the city of New York, 1807 by Isaac Riley, from Samuel Latham Mitchill, *The Picture of New-York; or, The Traveler's Guide, through the Commercial Metropolis of the United States* (New York: I. Riley and Co., 1807).
SOURCE: The Huntington Library, San Marino, California, RB 6607.

time it petered out at the end of October, the fever had killed at least 800 people, judging from the tabulations of the deceased in the city's churches, and probably many more if we consider the anonymous dead deposited in the Potter's Field. Davis recalled the unceremonious disposal that awaited these unfortunate victims: "For the speedy removal of the dead, a hearse was provided, drawn by an horse, and attended by two men. . . . they usually brought a coffin and tarred sheet with them, in which the corpse was wrapped, put into the coffin, and drove off to Potter's Field . . . and entered without the attendance of a single connexion or friend to bemoan their loss." Recollections of the fevers that struck Philadelphia in 1793 and Baltimore in 1794 exacerbated the tension. "The recent sufferings of Philadelphia and Baltimore, occasioned a greater fear of infectious fevers than would otherwise . . . have existed," Davis remarked.[14] Reflecting on the consecutive epidemics eroded confidence that the fever might disappear as suddenly as it came. Already on September 7, 1795, Dr. Amasa Dingley concluded that the prevailing fever was the same one that had devastated the country for the last three seasons, and that the disease would likely recur again and again.[15]

The fever particularly troubled Noah Webster, whose thoughts in the fall of 1795 turned to the dire consequences of recurring epidemics and continuing medical disagreements. In a "Circular" published in October 1795 and addressed to physicians all over the United States, Webster aptly depicted the crisis dawning over the United States. "As a malignant fever, has, for three summers past raged in different parts of the United States, and proved fatal to great numbers of our fellow-citizens, and extremely prejudicial to the Commerce of the Country, it becomes highly important to take such efficacious steps as human wisdom can devise to prevent the introduction, arrest the progress, or mitigate the severity of such a serious calamity." The very "happiness of families and the general prosperity of the country" hung in the balance. As a remedy, in his "Circular," Webster urged his colleagues to send him the appropriate "facts" of each occurrence. If collected together and published, Webster predicted, such information would produce "universal conviction" of the fever's cause.[16]

Webster's interest in yellow fever exposes an unfamiliar side of the well-known figure. Born in 1758 in the small Connecticut town of West Hartford, Webster attended Yale College under the presidency of Ezra Stiles. At Yale, Webster attended Stiles's lectures on astronomy, mathematics, and science, and he otherwise adopted the Scottish Enlightenment's commitments to empirical study and common-sense reasoning.[17] After graduating in 1778, Webster studied law for a time, passed the bar in 1781, but then abruptly abandoned the legal field in order to follow other pursuits. In the 1780s, he made a name for himself as a writer of popular textbooks, including the much-celebrated *American Spelling Book* (1783) and *The Prompter* (1791), a collection of pithy, moralistic sayings, which bore the revealing subtitle *A Commentary on Common Sayings and Subjects, which Are Full of Common Sense, the Best Sense in the World*. In 1791, Webster moved to New York, where, at the behest of Alexander Hamilton and other federalist leaders, he began to publish the *American Minerva*, a newspaper organ of the federalist creed, which he edited until 1797.[18]

With the help of his friends in New York, Webster's *Circular* bore some fruit. Published in its final form in late 1796, only months after the initial Circular, the *Collection of Papers on the Subject of Bilious Fevers* contained eight papers from observers in the five major seaport towns that yellow fever had struck—Philadelphia, New York, Baltimore, New Haven, and Norfolk. In true Hippocratic fashion, the essays featured historical narratives of the rise and fall of yellow fever in each location where it occurred, as well as detailed observations of the weather and the conditions of the cities. The effort amounted to no less than a complete history of yellow fever in the United States up to its publication. As one of the writers in the anthology declared, "By thus placing within so small a compass, the practical experience of those who have had an opportunity of treating and being acquainted with that disease, much useful knowledge of the subject may be disseminated, and the general good of mankind promoted."[19] By anyone's estimation, Webster had delivered a success.

The *Collection* might have been considered a resounding success except for one particularly bedeviling fact—Webster's hope for a "universal conviction" about the cause of yellow fever never materialized. Despite the preponderance of papers that attributed the sickness to local sources, one correspondent, Dr. Eneas Monson, marshaled compelling evidence that the fever that struck New Haven in 1794 originated from contagion. On June 15, 1794, Monson visited a patient, the eight-year-old daughter of Elias Gorham, who had yellow fever. Upon arriving, Monson reported observing "her countenance flushed with a deep red colour; her eyes were dull, and highly inflamed; she had violent pain in her head, back, and limbs; nausea, and frequent vomiting; obstinate costiveness; a quick, full, hard, throbbing pulse; her skin was hot and dry; and her tongue covered with a thick white fur." On the sixteenth, the girl's symptoms ceased—"her pain and distress suddenly abated," Monson declared. But, then just as suddenly, she relapsed, vomiting "up matter resembling coffee-grounds." The next day, the daughter of Elias Gorham died.[20]

Dr. Monson inquired into the girl's activities leading up to her illness. The girl's mother told Monson that she had been visiting with her aunt, the wife of Isaac Gorham, who lived near the "*Long-Wharf*" in New Haven, about three-quarters of a mile from Elias Gorham's house. The deceased girl's aunt, named Polly Gorham, had also taken the fever and died, succumbing on the fifteenth. Further investigation revealed that an infected vessel from "Martinico," piloted by a Captain Truman, had arrived in New Haven near the beginning of June and then docked at the wharf "within a few rods of Isaac Gorham's house."[21] Captain Truman then unloaded a "chest of clothes" belonging to a sailor who had died of yellow fever, and opened it in the presence of three observers. All three contracted yellow fever and died. Meanwhile, by the time Monson began to put the pieces together, Elias Gorham's wife, as well as Isaac Gorham's three children, had also contracted yellow fever. Isaac Gorham lost his infant and young son. In the city of New Haven, sixty-four people died of yellow fever.[22]

The evidence befuddled Webster. Though he would later denounce the "common" theory of contagion as "ill-founded," at the time of the publication of the *Collection*, Webster would not go quite so far. Equally impressed by the evidence of both theories, Webster staked out a ground somewhere in between. "One thing may safely be averred," he hedged, "whether imported, or generated by local causes in our own country, the epidemic influence and destructive effects of this malignant bilious fever, are greatly increased by local causes, which are wholly within the command of human power."[23] As Webster saw it, though the proximate cause, or spark, of the disease might be miasma or contagion—the facts alone did not permit firm conclusions—yellow fever nevertheless took its power, or its "epidemic influence," from particular features of the environment, especially the conditions of the afflicted cities, which were rife with noxious, disease-causing substances.

Webster gave no indication that he knew it at the time, but the so-called "epidemic influence" of the air had puzzled medical inquirers since the time of Hippocrates. Writing in the seventeenth century, the celebrated English physician Thomas Sydenham called this elusive state the "epidemic constitution," a term that Webster would soon adopt in his own writings. According to Sydenham, though diseases arose from their proximate causes (miasmas or contagions), their occurrences as epidemics depended on the "epidemic constitution." It was the force that transformed ordinary diseases into epidemics. But Sydenham's "epidemic constitution" was also fundamentally inscrutable. He believed that it resided in qualities of the air and environment, but ultimately its existence was known from its effects, not from investigation of its nature or properties. Nevertheless, the concept enabled medical inquirers to account for fundamental questions (the same ones that troubled the fever investigators): Why did the same disease sometimes strike as an epidemic and sometimes not? Why did epidemics sometimes fail to occur when their causes obviously existed? Thus, while the *Collection* may have failed to produce unanimous opinion about the proximate cause, it did get people to think more historically about the problem of yellow fever. The question as Webster asked it was not "what caused yellow fever," per se, but "why did yellow fever suddenly arrive in the United States and afflict its seaports for three consecutive years, when it had been almost entirely absent before?"[24]

Impressed by Webster's *Collection*, but dismayed by its failure to settle the controversy, a trio of medical thinkers in New York—the professor of chemistry at Columbia College, Samuel Latham Mitchill, along with Edward Miller and Elihu Hubbard Smith—made preparations for an even more ambitious historical project. Thrust together in New York's burgeoning yet intimate intellectual scene, the unlikely threesome made a motley group from the start. Only Mitchill was a native New Yorker; he was born in North Hempstead, Long Island, in 1764. He apprenticed for five years under the tutelage of his uncle, Dr. Samuel Latham, then with Dr. Samuel Bard, before traveling to the University of Edinburgh, where he studied with the aging William Cullen (who would die in 1790), graduating with a medical degree. Upon returning from Edinburgh in 1787, Mitchill took over the professorship of chemistry at Columbia College. Sometime after 1793, he met Smith, an eager young poet and budding physician. It was probably Smith who introduced Mitchill to Miller, a doctor originally from Dover, Delaware. Like Mitchill, Miller apprenticed for several years in his hometown before moving to Philadelphia, where he earned his medical degree from the University of Pennsylvania in 1785. After moving to New York, Miller began to pay regular visits to Friendly Club, of which Smith was a member and Mitchill a regular attendee.[25] Their collaboration would lead to one of the most influential intellectual projects of the early republic.

Smith seems to have hatched the idea in the summer of 1796. In a diary entry of August 11, having learned that Webster would no longer collect evidence about yellow fever, Smith mused over "taking it up myself . . . and publishing an annual volume; the principal object of which will be the preserving & collecting of the materials for a History of the Diseases of America."[26] In early 1797, Smith and his colleagues drafted a proposal for a vast, collaborative, fact-gathering enterprise, which, they hoped, would eventually result in a history of disease in the United States. Addressed to physicians all over the country, the proposal called for descriptive information regarding the occurrences of disease. The writers asked for five types of information: "Histories of such diseases as reign in your particular places of residence," "Histories of such diseases as appear among *Domestic Animals*," "Accounts of *Insects*," "Histories of the progress and condition of *Vegetation*," and "The state of the *Atmosphere*."[27] Like other historical efforts in the early republic, the trio's project celebrated the exceptionalism of America. Unlike European countries, the United States possessed "extensive territory" and a great variety of soils, climates, and peoples. Besides enabling them to discern the causes of diseases, the uniqueness and diversity of the United States would help the trio discern the effects of "gradual and rapid changes in the face of a country," and the influence of the "savage, civilized, and intermediate states of society." Once published, the authors confidently predicted, the collection "could scarcely fail, even in a few years, of leading us to a near view of the origin and causes of general, febrile diseases" and to "the discovery of what situations, climates, and seasons, most favoured their production."[28]

In their most essential details, both the trio's proposal and Webster's *Collection* conformed to the latest, most sophisticated ways of scientific knowledge. In their proposal, the authors specifically linked their plans to broad changes in science that extended all the way back to the Scientific Revolution of the preceding century:

After a continued struggle of many centuries against the absurd systems of ancient physicians, and amid the difficulties repeatedly opposed to the progress of Medicine by modern hypotheses scarcely less preposterous, it has length become established as a fundamental truth, that experiment must precede conjecture, and that facts are the only rational basis of theory. Philosophers are no longer permitted to descend from generals to particulars, shaping them according to preconceived notions of their intimate relations; but are expected to proceed by a rigid examination and cautious assemblage of particulars to every general inference.[29]

As scientific ideas gained ascendancy, the authors continued, "Collections of Histories and Observations . . . have gradually obtained a high consideration and authority in the schools of medicine." A substitute for experiment, history supplied the core evidence for empirical research; history furnished investigators with the materials for "rigid examination" and the "cautious assemblage of particulars." The projects of Webster and the New York trio also showcased common practices of Enlightenment science, with its emphasis on encyclopedic description and compilation of facts and observations, as seen in the works of Linnaeus and Buffon, as well Diderot and d'Alembert.

The authors of the "Address" never did publish the compilation they intended, at least not in any single history. Instead, the trio diverted their energies to a new project, a periodical publication that would feature news of disease in the United States as well as intelligence from scientific circles in Europe. In November 1796, Smith, along with Mitchill and Miller, finalized their plans for the *Medical Repository*, as the journal would be called.[30] The *Repository* satisfied the more pressing need for scientific intelligence, but the editors did not forget the commitment to the study of diseases that initially had set them on the path to the *Medical Repository*. The very first article of that inaugural issue was an essay titled "The Plague of Athens," written by one of the journal's founders, E. H. Smith. Only twenty-five years old at the time of publication, the native of Litchfield, Connecticut, and graduate of Yale College had great expectations for his work: "The study of the histories of those wide-wasting diseases which pass under the name of Epidemics . . . is calculated to excite a suspicion, that they all have one common origin." He continued, "Should a minute inquiry into every thing which relates to these pestilential maladies justify such a suspicion, we should, probably, discover their hitherto hidden cause, and be enabled to prevent its future operation." Smith hoped eventually to expand his examination of Athens into a much larger work on the "History of Epidemics," but only completed one more essay, a brief piece on diseases of the Punic Wars, before his tragic death from yellow fever in 1798.[31]

Smith's "The Plague of Athens" recounted the terrifying events of the pestilence that befell Athens in the fifth century BC. Smith emphasized the circumstances attending the plague—the "facts" of the epidemic—but he realized that his study suffered from a crucial lack of documentation. Though Hippocrates and, much later, the Roman poet Lucretius had written about it, only Thucydides had offered an eyewitness account of the epidemic, and he never decisively stated what had caused it. Noting only that some Athenians thought it had originated from Ethiopia and others from a poisoned well, Thucydides left it to the Athenian citizens "whether physician or not" to decide for themselves. Besides, Smith found good reason to doubt the testimony of his sources. "Though no person will venture to question the sagacity of the Athenians, and the peculiar talents of Thucydides for observation, neither his countrymen, nor himself, can fairly be supposed to have examined this subject with philosophical precision."[32]

Smith circumvented these evidentiary obstacles and showed that Americans had much to learn from the Athenian plague. He supplemented his reading of Thucydides with a generous sampling of the *Philosophic Dissertations on the Greeks* (1793) by the Dutchman Cornelius de Pauw, "in the fidelity of whose quotations," he confessed, "it has been necessary to confide more than was to be desired." After examining the conditions of Athens during the epidemic, especially its deplorable sanitation, Smith concluded that the Athenian plague arose from local sources. Smith realized that the circumstances attending the Athenian plague bore striking resemblances to the situations in the American port cities during the yellow fever epidemics. He found that Athens during times of peace had approximately the same population as New York and Philadelphia, about 50,000 permanent inhabitants, and that both ancient *polis* and modern cities had narrow, crowded streets, and cramped housing, with numerous sources of filth and putrefying matter. Athens and the afflicted cities in the United States also shared nearly the same latitudes and thus, Smith thought, the same climates. Even the symptoms of the diseases in Athens and the United States appeared similar. Since the Athenian plague so clearly resembled the American yellow fever, and since it occurred in nearly the same conditions, he wrote, "We may be justified in declaring it to have been, in all essential particulars, the same disease." Smith concluded, "If local causes originated a pestilence in Athens, local causes may generate a Yellow Fever in Philadelphia and New-York."[33]

Appearing as the very first article in the nation's first and much anticipated scientific journal (the editors drummed up subscriptions from 266 people in 14 different states), Smith's "Plague of Athens" testified to the growing regard for historical studies of disease. It also flattered the aspirations of American intellectuals, who fancied themselves Athenians reborn, and clearly imagined their cities to be like Athens. The poet Elizabeth Graeme Ferguson once termed Philadelphia, with its flowering cultural and intellectual scene, the "Athens of North America." Her regular salon gatherings were even known as "Attic Evenings" (a reference to the region in Greece), and included regular attendees such as Benjamin Rush. But Smith's essay also exposed a problem. Appearing at the beginning of the fever season of 1797, when the disease killed more than 1,500 people in Philadelphia alone, it revealed that there were dangers to being too much like the ancients.[34]

Meanwhile, in the summer of 1797, as the editors of the *Medical Repository* savored their newfound success, the return of yellow fever to Philadelphia set off another round of commentary. As the fever raged, William Currie wrote a letter addressed "to the citizens of Philadelphia," published on August 21 in *Claypoole's American Daily Advertiser*, in which he argued that a ship called the *Arethusa* had introduced yellow fever to Philadelphia. The *Arethusa*, a slave vessel, had arrived in port on July 18 or 19, 1797, with eleven people on board, none of whom were sick. Soon after arriving, the "pilot" of the vessel took ill with what Currie deemed to be yellow fever. Only five days later, five more sailors lodging aboard the *Iris*, a ship docked next to the *Arethusa*, also came down with symptoms of yellow fever. Currie's suggestions incited a skirmish of sorts in Philadelphia's newspapers, with Benjamin Wynkoop, a printer, taking the side of localists and Currie defending his own theory. Eager for news of the fever in Philadelphia, journalists in New York also began to run the Currie-Wynkoop debate.

The public debates between Currie and Wynkoop attracted the attention of Webster, who once again entered into the yellow fever controversy, presenting himself to the public as an impartial arbiter of opinions. In a series of twenty-five letters addressed to Dr. Currie, which appeared in Webster's newspaper, the *New York Commercial Advertiser* (formerly *American Minerva*), from October 26 to December 20, 1797, Webster presented trenchant analyses of Currie's argument. In the first letters, Webster reiterated much of what he had already claimed about the cause of yellow fever:

"Whatever be the truth as to the *first introduction* of the plague, the communication and prevalence of it as an epidemic disease, depend wholly on the constitution of our own atmosphere."[35] This time, however, Webster searched history for evidence. He examined recent epidemics, such as the one in London in 1665, but he focused on the diseases of ancient Rome and, of course, Athens, about which "Dr. Elihu H. Smith, of this city, has lately written and published . . . a very ingenious and interesting comment." Indeed, Webster wrote, "It is remarkable how exactly the plague at Athens, four hundred and thirty years before the christian era, resembled the malignant fever, which annually rages in some of our sea ports." Webster also discovered similarities between yellow fever and numerous other diseases: "We are then in the same situation as to exposure to malignant epidemics, as the ancient Greeks, the Romans, and the modern Turks."[36]

As Webster delved deeper into historical study, he encountered more and more evidence that cast doubt on contagionism. "Scarcely an instance could be found," he scoffed, "in which the evidence of the propagation of disease, from imported infection, was sufficient to render the fact even probable." To the contrary, history revealed that epidemic diseases almost always arose from domestic sources of miasma: "Ancient plagues were not propagated by specific contagion, but originated in *bad air.*" Due largely to the insights from the past, Webster had completely abandoned contagionism and embraced localism by the time he published the *Brief History* in late 1799.[37]

Having determined that the "remote cause," meaning the physical material cause, of yellow fever arose from domestic sources, Webster devoted the *Brief History* to deciphering the epidemic constitution of the air, the key to his emerging picture of disease causation. The *Brief History* is above all things a comparative history of diseases, based on minute historical research from primary and secondary sources. These sources ranged from Thucydides, Livy, and Procopius, to Edward Gibbon's *The Decline and Fall of the Roman Empire* and Charles Rollin's *Ancient History,* to the travel accounts of Volney, the Baron de Tott, Cornelius de Pauw, and Dr. Alexander Russell.[38] Webster hoped that by correlating diseases and other conspicuous natural phenomena he might be able to discern the patterns that would reveal the cause of the epidemic constitution. Webster focused on the major climatic and environmental happenings. "I shall note, as I proceed," he began, "any extraordinary occurrence or phenomena in the physical world, as earthquakes, eruptions of volcanoes, appearance of comets, violent tempests, unusual seasons, and other singular events and circumstances, which may appear to be connected with pestilence, either as a cause or effect, or as the effect of a common cause."[39]

Yet Webster's intellectual endeavor was fraught with difficulties. For example, he could not get all the books he required. The problem revealed the deficiencies of the United States, which lacked many of the resources available in European centers of intellectual activity. "No man can find in this country *all* the books necessary for a complete examination of a historical or scientific subject," he lamented in the *Brief History*. Not even the vaunted collection at the Library Company of Philadelphia or the private collection at Harvard could satisfy Webster's desire for books. "The public libraries of New-York and New-Haven, tho very valuable, are deficient. Those of Harvard College and Philadelphia, are more extensive, but incomplete." The dearth of works in foreign languages particularly troubled Webster. Though he easily found the writings of the Romans and the Greeks, he could not find those from modern writers in Italy, Germany, Spain, and "the Baltic nations," leaving gaping holes in his research.[40]

In a complaint echoed by modern historians, Webster also criticized the narrow, biased foci of the historians of his own age. "Most modern writers appear to think every thing beneath their notice, except war and political intrigues." Even the most celebrated historians of his time—"Hume, Robertson, Smollet, Rapin and Gibbon"—virtually ignored the plagues that desolated mankind, making the retrospective identification of their causes extremely difficult. As Webster saw it, the deficiency sprang not from any methodological error or absence of appropriate source material, but from deep-seated flaws in the characters of modern historians and the purposes of modern histories:

> In respect to useful history, the ancient authors have the preference over the modern. Modern compilers appear to have written for fame or for money, rather than for the sake of unfolding or diffusing truth. Hence they have principally attended to those animated periods of the world, which were distinguished for great achievements; or those prominent events, a description of which would interest the passions of their readers. . . . Others appear to have undertaken historical compilation, solely or principally to support some preconceived system of government or religion; and have studied to bend the evidence of facts, to the accomplishment of that purpose.[41]

Webster also attacked the motives and superficialities of travel writers, who "pass from country to country; examine and describe a few external objects, such as cities, buildings, paintings and statues, but leave more useful subjects unexamined, and return home with a book of vulgar tales and errors."[42]

Though Webster generally valued the testimony of ancient authors, he also discovered flaws with their firsthand accounts. In trying to reconstruct the history of the fifth century BC, an eventful and calamitous time in the history of Western civilization, Webster drew from a panoply of classical writers, including Livy, Thucydides, Herodotus, and Dionysus of Halicarnassus. But he soon realized that historians seldom agreed, even on the most essential details, such as the chronology of events. Webster wrote abashedly, "We are sometimes embarrassed with the differences in the chronology of different authors."[43] The problem proved particularly vexing to Webster, whose identification of the cause of the epidemic constitution depended on precise knowledge about the relationships between plagues and other natural occurrences.

Despite the numerous problems he encountered in his research, by the end Webster willingly offered a firm, even striking conclusion. Based on his exhaustive survey of the sources, Webster found that the yellow fever epidemics of the United States, as well as all other plagues in the history of mankind, had been preceded by "violent agitations of the elements."[44] Such agitations included principally earthquakes, and volcanic eruptions, as well as droughts, severe winters, and epizootic diseases. As he moved through history, time and again Webster discovered that major environmental occurrences always preceded the appearances of pestilence. In his investigation of history from the time of Christ, Webster studied forty-seven separate occurrences of plague and for each one he found corresponding evidence of some remarkable natural phenomena, such as earthquakes or volcanic eruptions.[45] Tellingly, the plagues and natural phenomena did not have to be confined to the same geographic area, nor did they have to be aligned in a precise chronological sequence. For Webster, it was enough that an earthquake or eruption had occurred somewhere on the globe, and that perhaps a few years or even a decade had separated the two events. He expressed no concern that the last major earthquake in the United States occurred in 1783, ten years before yellow fever began to afflict the seaports, or that the last preceding volcanic eruption took place on Mt. Vesuvius in 1789.[46]

To be sure, Webster never claimed that the events preceding the appearances of diseases actually caused them. Rather, he argued that the two phenomena were both effects of another cause, the elusive epidemic constitution. "The evidence I have collected," Webster delightedly wrote to Benjamin Rush in February 1799, "will demonstrate that the great primary cause of epidemics, is a quality of the atmosphere, extending usually over one hemisphere, sometimes over the globe."[47] Unfortunately, Webster did not know what produced the primary cause, the epidemic constitution, though he speculated that it might be the "*invisible* operations of the electrical fluid," which he thought surrounded the earth at all times.[48] Confiding to Rush in a letter dated November 26, 1799, Webster admitted, "I more and more see the difficulty of reaching the cause. That the electrical principle is the agent, I am inclined to believe; indeed, I have not the power to resist the evidence of it; but by what combination with aerial substances, I am ignorant."[49]

Even more remarkably, Webster tried to correlate the perturbations in the electrical fluid, and thus the rise of the epidemic constitution, with observable astronomical phenomena. In fact, he devoted large sections of the *Brief History* to discussing the positions of the moon and planets, as well as the appearances of comets during times of pestilence.[50] Webster believed that the proximity of heavenly bodies exerted force on the earth, altering its proportions of "electrical fluid" and thus preparing it for epidemic disease. The same force also, in some fashion, produced the major natural disasters that accompanied epidemics. "It is not indeed unphilosophical to suppose," Webster claimed, "the several immense orbs that compose the solar system, to have an influence on each other by means of the great laws of attraction and repulsion." As an example, Webster cited the great influence of the moon and comets upon the earth's seasons and the oceans' tides, which philosophers had understood "as long ago as the days of Aristotle."[51]

Webster's theory of disease causation, with all of its intricacies, bold claims, and deductive leaps, rested on simple assumptions about the makeup of the world and the human ability to understand it. God had created a magnificent world filled with wonder, but he had also endowed humans with the rational capacity to examine it, to reach firm conclusions about it, and thereby to better their own conditions. This shared capacity, the common sense, enabled inquirers to sift through the scattered phenomena of nature and to find truth among the repeated concurrences of events. The common-sense epistemology had lured Webster into the historical study of disease in the first place—indeed, it undergirded the entire effort among the fever investigators—and it sustained his conclusions. Webster had shown that certain facts accompanied each and every occurrence of disease in human history, and, he believed, those facts would not deceive. Thus, even though Webster could not pinpoint the precise causal relationship among the facts, he nevertheless asserted that there was one.

If Webster's *Brief History* strained the limits of acceptable science, it also occupied an ambiguous position in early republican historiography. Like mainstream histories of the time, it cast history as a cyclical affair, and it was premised on the belief that reasoned, enlightened inquiry would enable inquirers to escape the cycles of the past. But unlike exceptionalist historians of the era, and despite his unflinching nationalism (the same spirit that impelled him to craft his American dictionary in the hopes that the American language would one day become the universal language), Webster could not depict America as a place apart. His analysis showed, to the contrary, that America was still a part of history. Though Webster would not have liked to admit it, his *Brief History* resembled the works of the "modern" historians he deplored—Hume, Robertson, and Gibbon—particularly with its universalizing or "cosmopolitan" emphasis. As scholars of eighteenth-century history have shown, European Enlightenment historians strove to unite disparate peoples of the Western world around a single story, usually one that focused on the progress of civilization and enlightenment. Webster too placed the United States and Europe into a single narrative—a narrative of disease that privileged the miasmatic theory and united all peoples and regions through their common victimization by the occult epidemic constitution.[52] That Webster would deviate so sharply from his own assumptions about America's uniqueness showed how desperate the controversy over yellow fever had become, and how far historical study defied his expectations.

Webster's contemporaries evidently held many of the same opinions, at least judging from their overwhelmingly positive reactions to the *Brief History*. Benjamin Rush gushed over some of the early drafts of the arguments that Webster would feature in the *Brief History*. "You have opened a mine of precious metals to the lovers of science and friends of humanity," Rush began in a letter of March 1798. "I have for some time past suspected that the malignant constitution of the atmosphere taken notice of by Hippocrates and Dr. Sydenham pervaded our whole globe, at the same time and that it was somehow produced by the influence of some part, or parts of the solar system upon our planet. But you have demonstrated that to be true which was only a floating idea in my mind."[53] According to Rush, Webster's historical work had actually illuminated a facet of a great mystery in the history of medicine. The epidemic constitution hovered over large parts of the earth, sometimes the whole planet, and it was intimately tied to the movements and positions of the heavenly bodies.

In their review of the *Brief History*, the editors of the *Medical Repository* added to the praise of Webster's work. "Whether we consider the extent and novelty of the design, the diligence, acuteness, and erudition displayed in the investigations, the ingeniousness and sagacity of the combinations, or the importance of the results; whether we consider the revolution of opinions which it tends to effect, or the range of inquiries which it is adapted to awaken." However one considered it, "In our judgment Mr. Webster has performed a great work."[54] The editors particularly praised Webster's methodology. Having assumed the "station of an historian," Webster's research and conclusions rested on a solid empirical foundation—the countless examples of the correlation between diseases and other natural phenomena. Consequently, Webster had produced "luminous and incontrovertible proof" of the epidemic constitution.[55] The editors also tempered their praise with some cautionary advice about the true meaning of Webster's conclusions. "It is requisite to know much more of nature," they admonished their readers, "before we can hope to ascertain the precise degree of connection or dependence of these occurrences upon one another." Webster, after all, had not demonstrated causality; he had only shown that a broad range of coincidences indicated—and, as he and others saw it, quite forcefully—that disease and other observable phenomena shared a relationship of some sort. Whether the phenomena caused the diseases, or whether they were only produced from another common cause, however, they did not know. In order to clarify these outstanding questions, the editors urged philosophers to gather "more numerous and more minute observations."[56]

At nearly same time as Webster published his *Brief History*, another inquirer, James Tytler, released his own history of disease, *A Treatise on the Plague and Yellow Fever*. Published by an obscure press in Salem, Massachusetts, the *Treatise* evidently took its contemporaries by surprise, unlike the highly anticipated work of Webster. The editors at the *Medical Repository* complained that they only received the volume months after its initial release.[57] Mystery also enshrouded its author, James Tytler, who had not written anything on yellow fever before the *Treatise*. In fact, he had only recently escaped from Scotland in 1793, where authorities brought him up on charges of sedition. While still in Scotland, Tytler had practiced medicine and surgery. He had even achieved some renown as the author of the medical portions of the *Encyclopedia Britannica* and the *System of Surgery* (1792), a practical guide for surgeons. Famously, he had also been the first Briton to pilot a hot air balloon, a feat that earned him an unimaginative nickname, James "Balloon" Tytler. After the release of the *Treatise*, he lived the rest of his life in obscurity in Salem, until his death in 1803.[58]

Like the *Brief History*, Tytler's *Treatise* drew from numerous sources in order to reconstruct the history of epidemic diseases and ascertain their causes. In the first section, Tytler followed the history of the true plague, the bubonic plague, from its first supposed appearance in the Bible, to the much more recent accounts of its activities in the "Turkish dominions," especially Constantinople and Egypt. Along the way, Tytler described the plagues that struck the Greek world during the Trojan and Peloponnesian Wars, as told by Homer and Thucydides, respectively. He discussed the plagues that beset the Roman Empire in the tumultuous third century AD, and the harrowing epidemic of Constantinople under the reign of Justinian. Tytler also covered the famous epidemics in London in 1665 and Marseilles in 1720, and, at the end of the book, he even appended short excerpts from some of the more famous plague narratives in history.

Tytler offered strikingly different conclusions from Webster. Upon examining the conditions that accompanied plagues in history, Tytler found no evidence to suggest that disease and the environment shared any relationship whatsoever. To the contrary, he discovered that similar diseases frequently arose from contrary conditions, and that similar conditions did not always produce diseases. Causal inconsistency, combined with the observations of others, convinced him that diseases, yellow fever included, arose from imported contagions, or particles of invisible matter that imparted sickness to people.[59]

But rather than stopping at that conclusion—a rather unsatisfying one, because it did not posit a first cause—Tytler proposed a new way of accounting for the advent of contagions. Actually, Tytler's theory was not new at all, but very old, perhaps the oldest theory about the advent of disease in human history. In short, Tytler simply asserted that God had sent diseases to humans as punishments for their sins. As he saw it, God had created each separate disease in order to punish specific acts of disobedience.[60] One could even trace these creations through history in some cases, provided that one possessed adequate evidence. According to Tytler, the Bible preserves a record of the very first outbreak of epidemic disease in human history, which occurred when David committed the sin of "numbering the people," or imposing a census on the kingdom of Israel. David's oversight implicated his pride and thus provoked the ire of God, who offered David three choices of punishment, of which David chose the plague.[61] Having come into being, the plague particles simply traveled around the world from that point forward. Tytler even pursued the contagion to its next destination, the camp of the Greek warriors outside Troy during the Trojan War. As he informed his readers, "David was nearly contemporary with the Trojan war." The accounts of two outbreaks from the Bible and the *Iliad* also shared many similarities:

> Both plagues were inflicted on the people for the sins of their kings; both were miraculous; the one continued three days the other nine. In both the Deity himself appeared: an angel brandished a drawn sword over Jerusalem; and Homer says, that, from the top of Olympus, Apollo shot his arrows into the Grecian camp. Lastly, both were stopped in a similar manner: David offered sacrifices to the true God; and Agamemnon returned Chryseis, to her father, the priest of Apollo, by whose prayers and sacrifices the plague was stopped.[62]

From the site of the Trojan War, the plague simply migrated to the other places it afflicted.[63]

Compared with contemporary histories, Tytler's *Treatise* illustrates divergent trajectories in late-eighteenth-century historical writing. Typical historians of the early American republic held that time operated in cycles and that human reason could apprehend those cycles and, hopefully, elude them. They stressed that humans were drivers of historical change because they wanted to believe that there was something unique about America and Americans; they wanted to believe that they could change the patterns. To the contrary, Tytler subordinated the world of humans and their decisions to a sacred history, with its more linear plotline, which placed God at both the beginning and the end. His history left little room for reason and enlightenment, for what were Americans to do about contagions that wandered around the globe? Still, Tytler's depiction of disease history did anticipate the nineteenth-century transition from circular to linear history, and the resurgence of a more overtly providential history, which cast westward-expanding Americans as the carriers of God's purpose and the fulfillers of historic destiny.[64]

Tytler's appeal to divine intervention provided an explanation for the major problem in the contagionist argument—the error of infinite regress. One could not trace diseases backwards infinitely, as his analysis of the historical record showed. Diseases came into being at specific moments in time, as punishments for specific acts of wrongdoing. Tytler, however, could only supply a terminus to the regress by positing an explanation that most natural philosophers at the end of the eighteenth century considered invalid. Claiming that God did it was no way to account for natural phenomena. As a result, Tytler's peers censured his *Treatise* and the arguments it contained. In the *Medical Repository*, for example, the localist editors dismissed Tytler's contention about the divine origin of disease, though they otherwise applauded his efforts at compilation.[65] Even more telling was the resounding silence from the contagionists, who never once appealed to his work or defended it.

The epidemic constitution explained the philosophical problems at the heart of localism and thus helped turn the debate in its favor. Localists, recall, always faced troubling philosophical questions: Why did yellow fever sometimes fail to occur in the appropriate environmental conditions? Why, moreover, did diseases such as yellow fever erupt in different places at the same time if they were tied to local environmental conditions? The epidemic constitution offered answers. In his dissertation on the cause of yellow fever, a medical student named William Chalwill defied contagionist critiques of localism by explaining that yellow fever occurred simultaneously in the different parts of the United States and the West Indies because the mysterious epidemic constitution hovered over both areas. Chalwill pointed to similar circumstances in the past: "Thucydides tells us in his account of the plague at Athens, that it prevailed at the same time in part of Ethiopia, in Egypt, in Libya, and over a considerable extent of the Persian dominions. Procopius and Evagrius tell us that the plague which broke out at Constantinople, during the reign of the Emperor Justinian, lasted fifty-two years, and spread its influence over the whole earth."[66]

In certain instances, the epidemic constitution tipped the balance in favor of localism. In his 1804 dissertation, a lengthy historical work that compared yellow fever to bubonic plague, the localist Phineas Jenks confessed that he once embraced contagionism. But, he continued, "My researches, and the mass of facts I have collected have convinced me that my opinions were erroneous, and that I had fallen into a popular mistake." In a transparent indictment of the contagionists' infinite regress, and the arguments of Tytler, Jenks wrote with a convert's incredulity about the people who still attributed plague to "an obscure, and . . . distant origin, and sanguinely presumed to say, that it has had an existence ever since, being constantly kept up by contagion." Jenks ridiculed the alleged divine origin of disease by comparing it to the ancient Egyptian notion that disease arose from "flying serpents."[67] On the other hand, he found that the epidemic constitution easily resolved the causal inconsistency at the heart of localism. Diseases such as yellow fever often failed to occur when their causes obviously existed because they lacked the assisting influence of the epidemic constitution. With Webster's epidemic constitution, localists could finally explain why causal inconsistency existed. To the contrary, as Jenks's dissertation shows, Tytler's historical arguments only showcased the absurdity of contagionism.

Similarly, in an oration delivered before the Academy of Medicine of Philadelphia on December 17, 1798, the young physician Charles Caldwell spoke at great length about the high value of the epidemic constitution. "Pestilence can become epidemic only, when aided by a concurrent constitution of Atmosphere," he declared. "The nature of that peculiar state of atmosphere, favourable to the propogation of pestilential diseases, has hitherto eluded the researches of philosophers." Still, Caldwell continued, "Reason and observation still declare it to be a quality resting, for the certainty of its existence, on evidence as substantial, as that which supports the great Newtonian principle, the gravitation of all terrestrial bodies." Caldwell's philosophical gaze moved to history and the events and phenomena surrounding the occurrences of disease. Borrowing from Webster's letters to Currie, for example, Caldwell wrote, "The records of medicine afford ample proof . . . of the frequent and striking coincidence . . . of pestilence and the occurrence of earth-quakes and eruptions from volcanoes." Caldwell also found proof of the epidemic constitution in the fact that certain insects abounded during times of disease. Speaking about the latest occurrence of yellow fever in Philadelphia—a terrible epidemic, which carried off more 3,500 hundred people—Caldwell connected the presence of "muskitos," which "were more than usually abundant," to the existence of the epidemic constitution. "From the well known circumstance, that muskitos uniformly abound at the same time, and in the same places, with epidemic bilious fevers, we are very fairly authorized to conclude, that these noxious insects depend, for their existence, on an insalubrious state and constitution of atmosphere."[68]

By contrast, the contagionists did not make use of the epidemic constitution. This was not because it was incompatible with contagionism—a contagious particle could theoretically activate in an epidemic constitution just as soon as a miasmatic particle. The mode of yellow fever's introduction and transmission never troubled the contagionists, who knew that contagious particles could spread quickly and easily, so long as foreign vessels introduced them into affected cities. Beginning in 1798, when Webster's arguments began to circulate through public papers and philosophical circles, William Currie began to mount his own counterattack against the concept and its champion. He chided Webster for his reliance on Sydenham, a "very erroneous philosopher," who wrote at a time when "philosopher was only beginning to emerge from gothic darkness, in which it had long been sunk."[69] Tellingly, Currie did not deny that an epidemic constitution existed; he simply denied that it was anything like Sydenham and Webster claimed it was: "All that can with propriety be understood by an epidemic constitution of the atmosphere, is, that it is rendered more capable of retaining miasmatic effluvia and contagious matters, at one time than another, or that it renders the human body more susceptible of contagious, as well as more liable to febrile diseases from other exciting causes, at one time than another."[70]

The tone of Currie's rebuttals grew more desperate as Webster's *Brief History* grew more popular. In his treatise on the 1799 epidemic in Philadelphia, he wrote unabashedly about the epidemic constitution: "The doctrine of Mr. Webster on this subject, notwithstanding his elaborate researches, appears . . . to be as much the creature of imagination as the tales of the fairies."[71] Webster's "creature" had indeed taken on a life of its own, and it threatened to return science to an age of "gothic darkness." Currie thus urged readers to notice that Webster's theory failed to marshal the experimental evidence that it needed. "Till those gentlemen subject the atmosphere to eudiometrical experiments, and demonstrate that such a constitution does exist, or that some material change has taken place, it cannot with justice or safety be considered as any thing more than the mere suggestions of fancy, and deserves no more respect than the visionary opinions which prevailed in the dark ages of Gothic ignorance, when the conjunction or opposition of certain planets were believed to be the cause of the plague."[72] Currie's critique dismissed comets and constellations, earthquakes and eruptions as illegitimate scientific evidence. Currie's reference to eudiometrical experiment also suggested that the cause of yellow fever's recent virulence might be rooted somehow in the chemical composition of the air.

Criticisms of Webster and the epidemic constitution echoed from beyond the Atlantic as well. In an 1801 letter addressed to the College of Physicians, bastion of Currie (his "intelligent and steady" colleague) and the contagionists, the English physician and surgeon John Haygarth offered his thoughts on the waywardness of the localists. A fellow of the Royal Society with a statistical and experimental bent, Haygarth had made a name for himself as a cofounder of the Smallpox Society of Chester and advocate for inoculation.[73] He had been following the news about yellow fever since its first appearance in Philadelphia and the West Indies in 1793. By 1801, Haygarth had acquired an intimate familiarity with the contours of the debate, and he had also forged close personal ties with key participants, such as Colin Chisholm, with whom he corresponded regularly. Formerly the physician to His Majesty's Ordnance in Grenada, Chisholm personally experienced the yellow fever epidemics of 1793 and 1794, and he later composed a lengthy treatise in which he argued that the contagious disease had been imported from Boullam. The two exchanged ideas and information about yellow fever for the rest of the epidemic period.[74]

In his letter, Haygarth swiped broadly at the Academy of Medicine, Rush, and the localists, but he reserved his harshest criticisms for Webster and Caldwell for their "whimsical and irrational opinions" about the epidemic constitution. Haygarth turned a localist critique against them. Loose coincidences did not establish causation, he warned: "The question of cause and effect is in many instances of disease difficult to ascertain. In most cases we have nothing to direct our judgment except the close connection of place and time."[75] Besides provoking the localist indignation, Haygarth's condescending skepticism channeled David Hume and his reduction of causality to events conjoined in time and space. Whatever its philosophical implications, the comment showed his belief that coincidences of asteroids, volcanoes, earthquakes, and disease outbreaks did not reveal causation, much less the existence of an elusive epidemic constitution.

Haygarth's invocation of causality rightly suggested that historical analyses of disease reflected deep philosophical assumptions. Webster's epidemic constitution may have failed to meet Humean standards of causality, but that did not bother Caldwell, or Rush, or Mitchill and Miller, Jenks, or a host of others, who found that it jibed with common sense, and in fact celebrated the concept because it rescued "conjunctions of events" from the morass of Humean epistemology and elevated them to the status of cause and effect. On the whole, investigators' appeals to history rested on the belief, grounded in common-sense epistemology, that such facts and observations could and *would* illuminate the causes of diseases.

Still, despite the localists' confidence in history, inquirers such as E. H. Smith and Webster undermined its epistemic value. In the course of investigating the causes of yellow fever, Webster completely rejected the testimonies of all contagionists, impeached the integrity of modern historians, and questioned the purport of modern history. In his "Plague of Athens," Smith even impugned the philosophical precision of Thucydides (his only eyewitness!). And, according to Webster, even the much-esteemed ancient historians provided sloppy, inaccurate chronologies and left gaping holes over large swaths of time in the historical record. The localists were beginning to fall into another kind of regress, one in which more facts and more observations continually held out, but never fulfilled, the promise of solving the mystery of yellow fever. By the end of the fever period, it would have been hard for proponents of disease history such as Webster and Mitchill to view the promise of historiography as sanguinely as they had at the beginning.[76]

Worse still, if Webster's epidemic constitution did exist, it meant at best that early republicans could only hope to avoid its effects, but they might never be able to eliminate them. Webster channeled the anxieties it conjured into a critique of cities, with their small houses and narrow streets. The problem as he saw it extended from a mindless imitation of Old World urban design. "The ancient construction of London cost that city nearly *two hundred thousand* lives in one century; and Cairo and Constantinople probably lose more than that number, every half century." He continued, "I firmly believe . . . that a *perseverence in our present mode of building cities, will doom them all to the same fate.*" Were Americans as independent as he hoped, they might have avoided the troubles of yellow fever, Webster indicated. But then, in a surprising admission, he acknowledged the United States' hopelessly inextricable position in history: "The period of general contagion may subside, and intervals of more general public health, may be expected," he began, "but the melancholy period of epidemics will often recur—and as the plague, in all its shapes, is the offspring of *causes*, mankind, wherever those causes exist, are destined to be afflicted." There was a tragic irony to Webster's endeavor—drawn to the study of the past out of the hope that it would rescue American cities from yellow fever and redeem faith in American exceptionalism, Webster had to confess that his historical effort both revealed the limitations of enlightened inquiry and threatened the exceptional status of America.[77]

Webster's dire prediction only underscored a fear that had been growing since the beginnings of yellow fever in 1793, when investigators first contemplated the historical meaning of their fever. History held up a mirror to the early republicans, and they did not like what they saw reflected back at them. Writing in the wake of the devastating fevers of 1798, the worst of the epidemic period, Mitchill provided ample testimony to this historical anxiety. "The experience of the inhabitants of ancient Rome, London, and indeed of most large and populous cities in Europe and Asia, have in the progress of their settlement, suffered excessively from mortal epidemics." He continued, "New-York, and some other cities and towns of North America, are beginning to suffer what other cities and towns in ancient and modern times have undergone before them." Yellow fever suggested that even the United States, the grand republican experiment, the *novus ordo seclorum*, might surrender to the same slow and inevitable historical processes it was expected to avoid.[78]

History did not last long as a tool in the investigation of disease, nor was the epidemic constitution vindicated by future scientific advances in the study of diseases. The main problem that the epidemic constitution was supposed to solve—the randomness or inconsistency of disease occurrences— once again emerged as a point of contention for those who argued for the local origins of diseases and it remained so until the microbiological breakthroughs of the nineteenth century.[79] Nevertheless, for the time, history and the product of historical study, the epidemic constitution, enjoyed wide popularity, and thus by looking at them closely we can begin to see why it was that localism commanded such respect among increasing numbers of people. History allowed the inquirers to imagine their own yellow fever epidemics as part of sweeping changes taking place over the entire surface of the earth; changes that certainly had some relationship with the "violent agitations of the earth" and the movements of the planets, though the localists knew not what; and changes, furthermore, that one could follow around in the historical record and detect in famous periods of epidemic disease. Perhaps, then, yellow fever only reminded them of what they probably already knew, but did not want to admit, as the nineteenth century dawned over the United States—that the world was in a period of tumult, and that their own places in it might only be marginally more secure than before. If that.

THREE

"Nature Is the Great Experimenter"

> Men who have been much taught, are apt to be deficient in the sense of the present fact; they do not see, in the facts which they are called upon to deal with, what is really there, but what they have been taught to expect.[1]

JOHN STUART MILL, *1869*

In his *Nomenclature of the New Chemistry*, a book published in New York in 1794, Samuel Latham Mitchill reflected favorably on the recent history of chemistry, the field of inquiry devoted to the study of matter. "Until quite a modern period," he wrote, "Chemistry consisted of little more than scattered facts, imperfectly understood, and badly arranged." The field, as Mitchill understood it, had long suffered under the defective guidance of the alchemists and their illusory quest for the secrets that would enable them to turn base metals into gold or produce elixirs of everlasting youth and beauty. "Cultivated for a long time by vain and visionary searchers for the philosopher's stone and the universal medicine, it partook, as might be expected, in no small degree of the uncouthness and barbarism of its patrons." Happily, change was afoot. Due largely to the work of "Lavoisier and some other French gentlemen," the field had only recently undergone a kind of transformation. These chemists had produced results that were "of a kind so novel and unheard-of, that it became necessary to invent original names to express them." As Mitchill opined in his paean to chemistry, "So many alterations, additions and improvements have been made, that it may almost be said to have undergone a transmutation."[2]

The "transmutation" to which Mitchill referred constituted the break-through for which Marcellin Berthelot invented the more lasting term *la révolution chimique*, or "the chemical revolution."[3] Widely credited to the pioneering research of Antoine-Laurent Lavoisier, the chemical revolution produced what Thomas Kuhn called a paradigm shift—a fundamental conceptual reorientation toward the natural world. Kuhn likened a paradigm to a type of vision or lens that refracts what the natural philosopher sees in his or her world. "When paradigms change, the world itself changes with them," he wrote.[4] After the chemical revolution, natural philosophers quite literally visualized the world in a whole new manner. It fostered an entirely new way of conceptualizing the elemental construction of nature—the very components of matter that made up the physical world, from the largest geological formations of the visible world, to the smallest aeriform particles that composed the invisible world of airs and gases.

As the epidemics persisted, the fever investigators increasingly turned to the new chemistry for concrete, empirically based answers to their questions about the cause, nature, and transmission of yellow fever. The chemical revolution emerged directly from experimental observations; thus, it satisfied the epistemology that demanded facts and observations. With its emphasis on the composition of air, the heretofore little-understood element of nature, chemistry also offered a particularly powerful way to visualize the invisible disease-causing agent; to see, as it were, the particles floating about in an invisible aeriform state. More importantly, it allowed the investigators to locate the creation of the disease-causing matter from the various interactions, mixtures, and exchanges that characterized normal chemical reactions. The new chemistry inspired hope that they might finally solve the mystery of yellow fever and of infectious diseases more generally.

As Mitchill was a leading champion and practitioner of the new chemistry, his career offers an excellent vantage on the influence of the chemical revolution on the American yellow fever investigators. Mitchill first studied chemistry under the illustrious Joseph Black at the University of Edinburgh, and he also visited Paris in the mid-1780s, where he met Lavoisier.[5] When he returned to New York, he became the professor of chemistry and natural history at Columbia College, and he immediately instituted Lavoisier's system of chemistry as part of the curriculum. Two years later, he published the *Nomenclature of the New Chemistry*, intended for use as a textbook. Under his stewardship, the *Medical Repository* carried numerous articles, reviews, and news pieces about the chemical breakthroughs in Europe and the United States.[6] Mitchill matched his efforts to disseminate knowledge of the new chemistry with an attempt to apply it in a treatise on chemical origins of yellow fever, the *Remarks on the Gaseous Oxyd of Azote*, in which he argued that the material cause of the disease was actually a gas called septon, which emerged from putrefaction.

Hailed as a groundbreaking inquiry into the causes of disease, Mitchill's *Remarks* and its highly speculative conclusions also showcase the peculiarity of the investigators' uses of Lavoisierian chemistry. Though they celebrated its potential, the American chemists who brought Lavoisierian chemistry to bear on the problem of yellow fever never quite employed it as its French creators intended. The chemistry of Lavoisier represented the most rigidly empirical form of natural inquiry, based solely on the facts that emerged from experimental observation. It required self-discipline and skepticism. But with few exceptions, the chemical investigators did not perform experiments in order to their reach conclusions. Instead they relied on logical deductions, imaginative leaps, and syntheses of past research. Their inquiries were fanciful and reverential—more contemplations of a kind of chemical sublime than systematic chemical treatises. Excitement, passion, and hope more than experimental caution and rigor defined the unique American experience with chemistry.

The investigators' chemical studies recast the story of the reception of Lavoisierian chemistry in the early republic. Historians have focused overwhelming on the ideas and practices that Americans borrowed from the French, though they have noticed that American chemists lacked the same record of success as their European counterparts. "At the practical level there was more lip service than actual achievement in the application of chemistry," John C. Greene wrote in his survey of American science during the "age of Jefferson."[7] As Greene and others cast it, the shortcomings of American chemists sprang, not from any conceptual or intellectual differences, but from basic institutional deficiencies. That conclusion itself revealed important assumptions—that American chemists shared the same outlook as the French, and that they would think and act like them if they were able; and that chemistry itself was like any other commodity that could be imported and applied to scientific problems, provided that the basic material requirements were present.[8] To the contrary, the fever investigators' unique take on Lavoisierian chemistry (and, indeed, their lack of achievements in hindsight) reflected deep philosophical differences that distinguished them from the French, whose predilections toward heretical philosophies soured Americans' opinions in the era of the French Revolution. Believing that with common sense they could grasp the chemical designs of nature, investigators proceeded without the experimental evidence demanded by Lavoisier and his colleagues. In their hands, chemistry revealed the mysteries of yellow fever and rescued science from the clutches of heretical philosophers, who threatened to reduce nature to mechanism and science to a cold, passionless pursuit. The chemical ferment of the yellow fever years showcased the formative influence of common sense, and it anticipated strains of romanticism in American science.

Chemistry also accentuated the differences between localists and contagionists. Being more amenable to common-sense inquiries, the localists indulged more freely in chemical contemplations. By contrast, the sole contagionist to prosecute extensive chemical studies did so with all the austere experimental rigor that Lavoisier himself demanded. In the context of the yellow fever debate, the new chemistry allowed localists to depict the material cause of the yellow fever as small aerial particles, and it offered ways of imagining the putrefactive processes that produced them. The new chemistry also enabled them to place the creation of the disease-causing matter in nature, to discern its emergence from the cycles of life and death that defined all natural operations. The story of chemistry during the epidemic period thus helps us understand the ultimate triumph of localism in the early republic.

Lavoisier's revolution offered three major contributions to the study of chemistry. First, Lavoisier showed that oxygen was the true agent in combustion, respiration, and calcination; in doing so, he did away with "phlogiston," a nonexistent substance believed to be present in bodies capable of combustion and the key ingredient in the paradigm for understanding essential chemical processes. Second, Lavoisier's research reconceptualized the elemental construction of the natural world by showing that a finite number of irreducible substances (oxygen, carbon, nitrogen, etc.) composed all matter. He showed that these irreducible substances, as well as many of the compounds they formed, could exist as gas, liquid, or solid (depending on temperature), for they represented states of matter, *not* elements. In a world still in the grips of the four-element theory of matter, this particular revelation considerably broadened chemists' appreciations for the range of possibility inherent in even the simplest chemical compounds. Third, Lavoisier and his colleagues reworked all these irreducible substances and their combinations into a comprehensive nomenclature that regularized the language of chemistry, and gave chemical structures the names they still bear today.[9]

Lavoisier's chemistry steadily won over erstwhile supporters of the phlogiston theory. Lavoisier himself labored tirelessly to advertise his theories in theatrical public presentations, papers, and published texts. He quickly converted a number of his French colleagues, such as Louis-Bernard Guyton de Morveau, Claude Berthollet, and Antoine Fourcroy, with whom he published the *Méthode de Nomenclature Chimique* (1787). Soon chemists across Europe were embracing Lavoisier's chemical system. Translations of his key works appeared in numerous languages, and his theories entered the curricula of the major European universities. Only a handful of English chemists initially resisted the Lavoisierian system, and even they soon capitulated, all except the stubborn Joseph Priestley, who defended the phlogiston theory until his death in the United States in 1804.[10]

American intellectuals also eagerly adopted the new chemistry. Chemists such as Mitchill at Columbia, John Maclean at Princeton, and James Woodhouse at the University of Pennsylvania all incorporated Lavoisierian chemistry into their curricula. Chemistry flourished in the American voluntary associations, such as the American Philosophical Society, whose members regularly presented chemical papers, and the Chemical Society of Philadelphia, founded in 1792 by Woodhouse. Chemists disseminated Lavoisier's ideas in published books, such as Woodhouse's *The Young Chemist's Pocket Companion* (1797), a moderately popular book full of simple experiments for aspiring chemists. Papers and news about chemical investigations also appeared in periodicals such as the *Proceedings of the American Philosophical Society* and the *Medical Repository*.[11]

Enthusiasm for the new chemistry focused on its centrality in natural operations. Chemistry was the study of matter, so it had consequences for all scientific pursuits. "Happy is our age, in which at last, we are acquainted with the elementary laws of existing bodies!," an émigré chemist named Felix Pascalis Ouvière exclaimed in an annual oration before the Chemical

Society in 1795. "Those laws which extend to all material objects, visible or invisible known or still concealed from our observation; those laws, the limits of which, we do not know, because we cannot trace where the limits of nature are to be marked; those laws form and constitute the science of Chemistry." So high was Ouvière's regard for the new chemistry that he proposed making it the root of the tree of knowledge, a favorite metaphor for the organization of human knowledge among Enlightenment philosophers.[12]

American doctors and chemists flatly expected Lavoisier's system to transform the studies of medicine and disease. According to Joseph Browne, a doctor from New York better known for his part in Aaron Burr's infamous colonization scheme, Lavoisier's chemistry formed "a new era in that science, which will most extensively influence the practice of medicine."[13] In his review of medicine in the eighteenth century, the doctor and historian David Ramsay ranked Lavoisierian chemistry among the most important innovations in eighteenth-century medicine, and he anticipated that its value to medicine would increase in the following century. Such sanguine predictions underscored an intimate and long-standing bond between chemistry and medicine. At least since the time of Paracelsus (1493–1541), healers struggled to devise chemical elixirs that prevented or remedied illnesses. Chemistry offered invaluable insights about the structures of living bodies, and the underlying causes of poor health. For most of the seventeenth and eighteenth centuries, chemistry survived in the curricula of major European universities through its role in medical education, and the most influential doctors of the period—Hermann Boerhaave, Georg Ernst Stahl, Joseph Black, William Cullen, and Benjamin Rush, among others—were also its leading chemists.[14]

In the disease-stricken United States, the new chemistry promised specifically to help resolve the mystery of the cause of yellow fever. The new chemistry particularly struck the investigators for its revelations about the air, the most mysterious state of matter, which had only recently been considered an element. Since the age of Hippocrates, medical thinkers had known that disease and air shared some special relationship. To those who encountered the ravages of malarial fevers, it seemed obvious that the moisture and heat of the air helped produce putrid miasmas. In the case of contagious diseases, it seemed clear too that contagious particles of matter floated *in* the air like dust or pollen. But the new chemistry transformed inquirers' understandings of what air actually was. "We are now able to analyze the atmospheric air," Browne declared, "which instead of being a simple element, is found to be a chemical compound." Benjamin De Witt, a doctor from New York, wrote similarly of the air: "Philosophers formerly imagined it to be a pure, simple, elementary fluid. . . . Modern discoveries, however, evince that it is by no means an elementary substance; but composed of different constituent parts, possessing chemical qualities, and having a very extensive and wonderful agency, in a great variety of the operations."[15] Since air represented a state of matter and not an elemental substance, investigators reasoned that any deleterious material could turn into a gas and harm people, just as the "airs" of pneumatic chemists had poisoned and killed the animals in their apparatus. With its inherent *expansibilité*, one of the recognized qualities of the gaseous state, such harmful gases could easily expand to fill the space of a room, or even more ominously, the atmosphere of a city.[16]

The new chemistry also altered investigators' perceptions of the processes that produced disease-causing miasmas. As chemists of the time understood it, when living matter lost its vital, vivifying energy (itself a mysterious and much-sought-after subject of chemical inquiry) and died, it underwent a process known as fermentation. During fermentation, the chemical compounds of animals and vegetables disintegrated, freeing elementary particles to form new combinations. Fermentation was not itself a harmful process. Chemists knew that the fermentation of sugary materials such as fruits and grains produced alcohols, and that was reason enough for many to celebrate the process. They also recognized it as the source of renewal and rebirth, since it decomposed dead materials and enabled them to reunite as new life. Chemists realized too that under proper environmental conditions, fermentation released the deleterious gases known as miasmas. But however harmful such miasmas might be, Lavoisierian chemistry offered a new way to investigate and explain the putrid fermentations that produced them, and thus to counteract or avoid their deadly effects.[17]

Lavoisier's chemistry came to the early republican fever investigators as a kind of revelation. It struck them with the realization that there was a hidden layer of nature, long concealed from view, in which the smallest and most elementary particles of matter interacted with the most enormous consequences. Chemistry exposed the world and everything in it as combinations of the same elementary substances. The irreducible substances of

Lavoisier fused together, dissolved, and joined again in different forms and combinations, but always according to natural laws. In certain forms, they comprised the poisons and diseases so deadly to mankind; in others, their antidotes and cures. The new chemistry seemed able to unravel the mysteries surrounding the diseases. With the guiding light of the new chemistry, Mitchill boldly predicted, "The doctrines of poisons and contagions will be as intelligible as those of digestion and respiration."[18]

Still, for all their admiration, American chemists did not accept the new chemistry unchanged. They altered it both consciously and unconsciously, and they applied it to problems such as yellow fever in ways that Lavoisier and his French colleagues never intended. Educated in the tradition of "philosophical chemistry" taught by Joseph Black at Edinburgh, chemists such as Mitchill did not embrace the experimental approach of the French chemists. As Lavoisier exhorted his readers in the *Traité Élémentaire de Chimie* (1789), "We ought in every instance, to submit our reasoning to the test of experiment, and never to search for truth but by the natural road of experiment and observation." According to him, if unaided by experiment, the imagination misled and deceived, since it "is ever wandering beyond the bounds of truth."[19] To the contrary, Black's chemistry was mainly didactic—it was taught in classrooms, not laboratories, to audiences of medical students, who would use it in their medical practices. John Robison, Black's successor at the University of Edinburgh, later wrote of his predecessor "that the vigour of mind which urges on to investigation, and delights in experiments, has never been a strong feature of the doctors Character, and that he has been satisfied with a just conception of the subject, and with applying to his own purpose the observations and experiments of others."[20] Much the same could be claimed of Black's students, such as Mitchill and Rush, and, by extension, those whom they trained at their respective colleges.

As the 1790s wore on, enthusiasm for the new chemistry diminished. By the time that Mitchill began to publish the *Medical Repository* in 1797, chemical pundits such as Mitchill and Woodhouse wrote vaguely of the "French chemists" and "French chemistry." Chemical pundits such as Mitchill and Woodhouse complained about the assertiveness of these "French chemists," who insisted that the key features of Lavoisierian chemistry were truths established upon undeniable facts. As Mitchill wrote with some incredulity and exasperation, "I learn, from France, that their leading chemists persist in considering the composition and decomposition of water established upon PROOF DEMONSTRATIVE."[21] With their precision instruments and scrupulous measurements, the French chemists aspired to a level of certainty reserved for the soundest knowledge, such as demonstrative mathematical proofs. Rather than offering their nomenclature for consumption, they seemed to impose it on the scientific community, a fact that struck many as an intrusion that underscored the French desire to commandeer chemical discourse. Adopting the role of peacemaker, Mitchill even tried to modify the controversial nomenclature, replacing terms such as "oxygen" and "hydrogen."[22] French assertiveness breached the ideal of enlightened philosophical discourse, by which truth claims were to be decided by the disinterested public, not by fiat. Writing in the *Medical Repository*, Priestley, the lone defender of the phlogiston theory, drew a parallel that must have occurred to others as well. Describing the chemical controversy as a "war," he referred to his enemy as "Buonaparte," and likened himself to "old Wurmser," an Austrian field marshal who lost a series of battles to Napoleon.[23]

Priestley may not have been an accurate representative of American chemists, but his comment rightly suggested that attitudes about French science were influenced by contemporary events in France, especially the revolution and its aftermath. The French Revolution both captivated and terrified American audiences, who followed the events first with hope then with horror. It began with the confiscation of Church property, the removal of Christian iconography from public places, and the substitution of the Cult of Reason for Catholicism. Irreligious measures quickly degenerated into chaos and mass murder, as atrocities such as the Reign of Terror and war in the Vendée soaked France in blood. By the end of the 1790s, the revolution lay in tatters. Many concerned Americans concluded that the effort fell prey to the vicious plots of atheists and deists, who maneuvered themselves into the positions of power, only to sabotage the revolutionary effort for some misguided, or perhaps evil, end.[24]

Whatever its scientific merits, Lavoisierian chemistry came from the same minds that engineered the French Revolution. In both the United States and Scotland, the new chemistry became linked to the revolution and the philosophies that undergirded it.[25] Lavoisier and his chief lieutenants—Morveau, Berthollet, and Fourcroy—were all engaged in revolutionary politics. Morveau was even the first president of the Committee of Public Safety, the very body that later consigned Lavoisier to death on the guillotine and perpetrated crimes against the French people. After the revolution, Morveau, Berthollet, and Fourcroy each occupied important posts in the Napoleonic government. Lavoisier and his lieutenants came of age in a Parisian milieu steeped in doctrines—materialism, determinism, deism, and atheism—that Americans deemed heretical and the same ones, they believed, that led to the chaos and bloodshed of the revolution. Lavoisier remained a Catholic, but his chemical disciples all embraced one or more of these doctrines. So too did the mathematician Pierre-Simon Laplace, one of Lavoisier's principal collaborators, who openly promoted a deterministic and materialistic view of the world that shocked and offended God-fearing Christians.[26]

Though Lavoisier avoided the extremes of Laplace, his severe experimental standards were rooted in a dangerous theory of the mind. The unflinching experimental methodology embodied in Lavoisier's exhortation revealed his adherence to the philosophy of the Abbé de Condillac, who totally rejected the existence of innate mental functions.[27] As Lavoisier wrote in the *Traité Élémentaire*:

> When we begin the study of any science, we are in a situation, respecting that science, similar to that of children; and the course by which we have to advance is precisely the same which Nature follows in the formation of their ideas. In a child, the idea is merely an effect produced by a sensation; and, in the same manner, in commencing the study of physical science, we ought to form no idea but what is a necessary consequence, and immediate effect, of an experiment or observation.[28]

Lavoisier's theory of the mind reduced natural philosophers to children, groping in the darkness, dependent totally on the scraps of information that they could gather from their sensory perceptions. It also placed him in a lineage of philosophers that started with the esteemed Locke, but went tragically wayward with the heretical Hume.

American chemists such as Rush and Mitchill roundly rejected the strict experimental program of Lavoisier and the mental theories that undergirded it. As Protestants, they celebrated the innate, God-given abilities of the mind, and they emphasized the design evident in nature. In one of his many treatises, Rush expressly acknowledged the advantages of "reason" over "solitary and mechanical experience," a reference to experimental science.[29] Mitchill likewise emphasized the limitations of experimental chemistry. Not only did it fail to answer many questions, it also diverted attention from more important tasks: "The excessive desire—the rage we may call it—of making experiments in rooms and work-shops, has considerably diminished the number of true observers of nature." He particularly lamented that the "rage" (a word that conjured visions of unrestrained passions and hysterical revolutionary fervor) of experimental chemistry diminished appreciation of God's works. "While engaged in minute operations with the imperfect instruments of human contrivance, within narrow and secluded apartments, the genius of our time too often suffers the great works carried on by the perfect machinery of the Deity, through the wide range of creation, to pass unnoticed or neglected." Natural inquirers did not need the artifice of experiment, because the design of the world revealed nature's secrets. "NATURE IS THE GREAT EXPERIMENTER," he exclaimed, "and what we most want are OBSERVERS AND INTERPRETERS OF HER PROCESSES. But man is vain and impatient, and will be busy."[30]

Mitchill's critique of experimental chemistry exposed the assumptions with which he and the investigators, at least the localists among them, approached the problem of yellow fever. Where the French demanded rigorous self-discipline, designed to regulate a wayward mind, the investigators celebrated the intuition and inspiration of lively, intelligent minds. Nature did not need to be tortured in some secluded laboratory to reveal its secrets; to the contrary, it would willingly give them, if it were asked. As they employed it, Lavoisierian chemistry was not so much a set of practices and knowledge as a broad perspective with which to interpret and understand the elemental design of the world. And in their minds, chemical processes were not isolated and disconnected phenomena to be studied in isolation; rather they were integral aspects of the grand design to be appreciated holistically, as part and parcel of the cycles that characterized all natural operations. In the chemical reactions that produced yellow fever, the investigators believed they saw the essential principles of life and death, regeneration and decay. Chemistry did not merely reveal the origins of yellow fever, it validated a view of the simplicity and wonder inherent in God's creation.

Rush offered his views on the chemical causes of yellow fever shortly after the first appearance of the disease. Better known for his medical theories, Rush, like all medical students at the University of Edinburgh, studied under Black, whose "ingenious" lectures, he would later recall, made chemistry his favorite subject.[31] In 1769, Rush became the professor of chemistry at the College of Philadelphia, a post he maintained until 1789. He also wrote *A Syllabus of a Course of Lectures on Chemistry* (1770), the first chemistry textbook published in the thirteen colonies. After a mild appearance of yellow fever in 1794, Rush turned his knowledge of chemistry to a troublesome question that nagged localists from the start—why did the fevers sometimes fail to occur when their causes were present? He answered simply, but revealingly, that a "preternatural quantity of oxygen in the atmosphere" constituted the principal predisposing cause of yellow fever and other malignant fevers. Oxygen, he thought, strained the body and predisposed it to contract the disease from pestilential miasma. Wasting epidemics occurred when the air was oversaturated with oxygen.[32]

Simple in appearance, Rush's proposition rested on a complex picture of how the body worked in health. "Life . . . is motion, sensation, and thought, excited by stimuli," Rush stated in his *Lecture on Animal Life*. "Good health," he continued, "consists in the unity or equable diffusion of this motion and sensation . . . and in the free and easy exercise of all the faculties of the mind."[33] According to his medical system, diseases arose from perturbations in the balanced operations of the body, and fevers specifically were produced by the overstimulation of the blood vessels. Thinking of the body holistically, as a balance of stimuli, Rush reasoned that oxygen, with its vivifying powers, might at times produce an overstimulation of the blood vessels and thus predispose the body to yellow fever. Others agreed. Alexander Hosack, an admirer of Rush and a student at Columbia, produced evidence that an unusual abundance of oxygen in the air around Martinique accounted for the fever's prevalence there.[34]

Though he deployed the terms of the new chemistry, Rush did not perform experiments to validate his speculations. To be sure, he claimed that "a want of eudiometrical instruments" prevented him from performing experiments. Eudiometers were devices that measured changes in the volumes of airs and were thought to provide an index of air's salubrity. Beginning in the 1770s, when Priestley and Alessandro Volta improved the instrument, European chemists from Henry Cavendish and Sir John Pringle to Jan Ingenhousz used eudiometry to measure and map the healthfulness of cities and countrysides known for their diseases.[35] Lack of equipment aside, Rush's approach to chemistry mocked the careful and cautious experimental sensitivity of eudiometrists. He acknowledged, but ignored, that Cavendish, the discoverer of hydrogen, had shown experimentally that the proportions of oxygen and nitrogen in atmospheric air were always and everywhere the same. Rush frankly admitted too that experiments would not alter his perceptions: "If it should be found hereafter, that no excess in the quantity of oxygen in the atmosphere takes place during the prevalence of malignant fevers, I shall still suspect it to be their predisposing cause."[36] His common sense would triumph over their measurements.

Fittingly, William Currie thought differently. Though he did not pursue the issue in nearly the same detail, Currie asserted that "febrile contagion," such as the one that produced yellow fever, thrived "in situations deprived of a due proportion of oxygen or pure air." He acknowledged too that contagions of all sorts required high temperatures to operate effectively. The suggestion highlighted Currie's amenability to basic localist ideas (a stance he had maintained since the beginning) just as it revealed his indebtedness to Lavoisierian concepts. Like the French chemists, Currie conceived of heat not as a form of energy (eighteenth-century natural philosophers lacked a paradigm for energy) but as a material entity, one of Lavoisier's irreducible substances called caloric. Caloric *was* heat conceived of as nearly weightless fluid. Thinking of heat as a medium through which material substances traveled, Currie reasoned that in hot conditions, contagious particles attached themselves to caloric, forming molecules that displaced the oxygenous parts of the air (hence his belief that febrile disease prevailed in areas "deprived of a due proportion of oxygen or pure air"). Outbreaks ended when cold temperatures and frost detached caloric, he explained, "letting the contagious particles fall to the ground."[37]

Currie's brevity left questions, but his tentative suggestion highlighted a growing fascination with the powers hidden in the air. Writing almost simultaneously, the chemical enthusiast and localist Browne offered a different and far more detailed explanation for how the lack of oxygen, which he called "animal vital air," caused yellow fever and incited massive epidemics. Like Rush (but rather unlike Currie), Browne premised his explanation on a comprehensive view of how the body worked in health. Since "animal vital air" animated the living organism, he concluded that its absence unraveled the organization of animal life, sinking the body into deadly fever. According to Browne, all life and all nature operated in cycles of death and rebirth, which recycled all of the parts of living organisms. "All and every part of creation, as well animate as inanimate," he claimed, "should be in a constant state of renovation, thro' a continual circulation of decompositions, and successive generations." Moreover, at all times, organisms inclined toward putrefaction, or the breakdown of their organization, but they were sustained by a principle he called "electric attraction, or affinity of composition." Life existed in perfect, complementary balance: Animals took in "animal vital air" and gave off "vegetable vital air," while vegetable life took in "vegetable vital air" and gave off "animal vital air."[38] The intake of these airs preserved bodies against the continual inclination toward putrefaction.

Occasionally the equilibrium among animal and vegetable airs went tragically out of balance. According to Browne, such imbalances produced fevers, including of course yellow fever. It started with the diminution of the animal vital air, which he thought occurred during the hot seasons, but then culminated in a vicious epidemiological cycle. The absence of animal vital air incited yellow fever in the human body, whose chemical affinities putrefied more quickly; the afflicted body, in turn, emitted greater quantities of vegetable air, which made it easier for others to develop yellow fever.[39] Putrefaction lay at the heart of yellow fever and other miasmatic diseases (that is why disease inquirers for centuries implicated fetid marshes, piles of garbage, corpses and carcasses). Browne's theory gave "bad air," or miasma, a distinct chemical identity and rooted that identity in the perspectives of the new chemistry.

Though the balances between life and decomposition, vegetable and animal, could spiral out of proportion, they nonetheless testified to the majesty of God's design. For Browne, the new chemistry illuminated the great wisdom that pervaded creation, ensuring the equilibrium among the parts of nature. Yellow fever only arose accidentally from the occasional perturbations in an otherwise perfect system of balances. These revelations inspired in Browne a sort of contentment with and resignation to the inevitable successions of combination and decomposition. For upon death, Browne ruminated, the particles that once composed the living being would more quickly form new combinations—"they will quit the solitary abode of the dead, and will again be immediately brought into an active state of existence; their first combinations will be to form vegetables, probably odorant plants and sweet scented flowers will owe their sweetness and beauty to the identical corpuscles that once animated our dearest friends and nearest relations, from whence through the medium of the atmosphere they will probably again constitute the rosy health and lively bloom of toasted beauty."[40] Even life and death emerged from the new chemistry as relative states of composition among substances.

Theories like those of Rush and Browne validated intuitive and deductive forms of inquiry, and they affirmed the vibrancy, wonder, and design of nature. They also highlighted growing dissatisfaction with the purely mechanical and experimental approaches. Such theories, in other words, suggest their authors' sympathy with the romantic movement then blossoming into full view in Germany. Romantic natural philosophers such as Friedrich Schelling and Alexander von Humboldt rejected the "cold philosophy" of mechanists and experimentalists. As Goethe wrote, "What she [nature] won't discover to your understanding / you can't extort from her with levers and screws." Romantics sought refuge from dangerous trends in the philosophy of Immanuel Kant, whose *Critique of Pure Reason* (1781) argued that the mind inherently possessed the machinery for ordering chaotic phenomena into rational experiences. Ultimately Kant's ideas offered a more complete and lasting alternative to Humean skepticism than the common-sense philosophies of Thomas Reid, James Beattie, and even Benjamin Rush. Liberated from doubt, romantics celebrated the majesty and intelligence that pervaded nature, which they viewed more as an organic whole than a collection of autonomous parts. Humboldt would come to use the term "cosmos" to describe this picture of nature as "a harmoniously unified network of integrally related parts."[41]

Americans themselves never had much contact with the Germans, and there is no reason to think that they consciously borrowed from them, but they did subscribe to some of the central notions.[42] The romantic impulse was imbedded in their natural history, in their fascination with the American landscape and their search for the untold powers that lay hidden in it.[43] It came through in Mitchill's claim that nature was "the great experimenter," and in his call for a type of broad-based, observational empiricism, through which nature was to be appreciated more than analyzed. It appeared too in Rush and Browne's awe in the face of nature's majesty, and especially in their insistence that the cause of yellow fever would fit holistically into natural operations. For whether yellow fever was brought on by an excess of oxygen or vegetable vital air, their ideas made sense only because they were premised on a view of nature as a "harmoniously unified network of integrally related parts." That was why disturbances in the elemental composition of air disrupted the healthy functioning of the body—air and body were bound together in a perfect balance. As the likes of Rush and Browne cast it, the harmony of the theory—its ability to fit plausibly into the apparent design of the material world—was itself proof of its correctness. Such formulations appealed to common-sense understandings of nature and natural inquiry.

Romantic sensibilities also came with outsiders. One such visitor was Felix Ouvière, who left an important mark on the yellow fever ferment. Born in France in 1750, Ouvière escaped to Philadelphia in 1793, one of the thousands of French-speaking refugees escaping the terrors of the Haitian and French Revolutions. He found a ready home in the United States, which suited his intellectual predilections. Unlike his Parisian chemical counterparts, Ouvière had been educated at the University of Montpelier, a center of medical vitalism, comparable to the Universities of Edinburgh and Pennsylvania. Montpelier professors specifically set their curricula against the dangerous mechanistic, atheistic, deterministic currents of thought then popular among the Parisian natural philosophers. Students at Montpelier learned the fabric of the human body could not be reduced to universal laws, and they were taught to look for disease in broad disharmonies of an otherwise balanced "body-economy."[44] At some point, Ouvière embraced the Lavoisierian system and dedicated himself to chemistry.

Ouvière's moment came in the fall of 1795, when the Medical Society of Connecticut sponsored an essay contest on the question: "What are the chymical properties of the effluvia of contagion of the epidemic of New-York, in the year 1795; what its mode of operation on the human body; and does said epidemic differ from the usual fevers of this country except in degree?" Ouvière's winning response was later published as the *Medico-Chymical Dissertations on the Causes of the Epidemic Called Yellow Fever*. Ouvière's chemical theory began with observations. Not his own, tellingly, but those of Dr. Jean Devèze, a French doctor, who dissected the corpses of several yellow fever victims during the 1793 epidemic in Philadelphia. The morbid appearance of the bodies revealed signs of tremendous internal damage. Drawing from Devèze's account, Ouvière reported: "There was found within the arterial system, only black and fetid blood, mostly rejected, in great quantities, through the cavities of several viscera, which were often mortified, detached, and floating in that corrupted matter." The heart itself seemed to have been spared the worst of the violence, but it "was of a pale colour, of a flabby consistency and as if it had been washed," and the vena cava contained many "white or red brown clots." The autopsies convinced Ouvière that the blood had lost its "aqueous particles," a phenomenon that he attributed to evaporation. He further surmised that evaporation released "aqueous gas" from the blood, disrupting the normal chemical combinations that comprised the healthy body.[45]

Ouvière's theory held that the evaporation of the blood, and thus the occurrence of yellow fever, depended ultimately on heat, or as he imagined it, caloric. "Caloric is a substance which penetrates thro' the pores of all bodies," Ouvière noted. He believed that caloric possessed a "dilating power" that enabled it to disunite chemical molecules and form new bonds with irreducible substances. During the transition to air, for example, caloric literally attached itself to the substances that composed solids and liquids and carried them away as gases. Ouvière concluded that in the hot summers, when air teemed with caloric, human blood could evaporate in its vessels, producing the deleterious "aqueous gas." Once unleashed, the "aqueous gas" raged through the body, undoing the chemical structures in the body's vital organs and impairing their functions. In different circumstances, the same gas would have slightly different effects, producing the other locally generated diseases with which Europeans and Americans had grown familiar.[46]

Ouvière's theory bore the trappings of Lavoisierian chemistry, but it owed much more to the power of his imagination. Ouvière attached to caloric a more elemental significance. Tellingly, the essay began with a quote from Ovid's *Metamorphoses*—*nec tantos corpus sustinet aestus ferventesque auras*—"nor can the body bear such great heat and such seething airs." The passage described the cataclysmic ride of Phaethon, son of Helios, the wielder of the chariot of the sun in Greek mythology. According to Ovid, in order to prove his paternity, Helios allowed his son to drive the fiery chariot, but Phaethon, who could not bear the "great heat and seething airs," lost control of the vehicle and it scorched the earth, destroying whole cities and peoples. The passage invited readers to compare caloric to the mythical power of fire, a giver of life and death, and it cast yellow fever as a manifestation of a timeless struggle against the elemental forces of the universe.[47]

Caloric enchanted Ouvière with its diverse functions in natural phenomena. "Caloric!" he exclaimed in a speech before the Chemical Society of Philadelphia, "Astonishing principle of destruction and life!" To appreciate its overwhelming might, he claimed, "It would be necessary to advert to mountains which it undermines, to the frightful craters it opens on their most elevated regions, and to the immense torrents of lava with which it inundates afterwards cities and empires." One could then follow the path of this "indestructible agent" as it "divides itself in various scintillating meteors, or by its sudden combination with air, forms the lightning, and by tremendous electrical detonations, spreads terror and devastation among mankind." Caloric mediated that narrow barrier between the tangible substances, fluids and solids, and the invisible gases of the world. It could "create as it were aerial and invisible bodies, among all the known substances you can enumerate and those that can be suspended in a gaseous state."[48]

Ouvière's picture of the cause of yellow fever rested on simple assumptions about the nature of life and death. Nurtured by the vitalists at Montpelier, Ouvière was taught to think that life was characterized by binaries—life was motion; death was stillness. As he understood it, animals were composed of a variety of Lavoisier's irreducible substances—hydrogen, carbon, nitrogen, phosphorus, and sulfur—kept in motion by some vital force, the principle of life. Normally, that balance remained intact—"the chymists have happily discovered and explained, that the pressure of the atmosphere settles by itself a necessary equilibrium between the dilating effect of the caloric, and the very attraction between the particles of a body." But the harmony only occasionally went out of balance and disrupted "that uninterrupted renewing which constitutes life."[49] Yellow fever represented a perturbation in the cycles that sustained life. There was something to Ouvière's allusion to the ride of Phaethon.

Loose claims to plausibility notwithstanding, Ouvière's particular theory of the cause of yellow fever never quite caught on. In the *Medical Repository*, the great disseminator of information, Mitchill wrote a critical review article about Ouvière's treatise. He struck right at the heart of Ouvière's argument—the notion that blood could evaporate into a gas. Has a liquid, Mitchill asked, ever been known to turn into a gas at a lower temperature than the "80th degree of Reaumur's scale?"[50] The question exposed a critical flaw in the argument, but Ouvière did not put up much of a fight. Indeed, as it turns out, he soon abandoned his theory in favor of the rival notion proposed by Mitchill.

Mitchill published his view of the chemical cause of yellow fever in a treatise called *Remarks on the Gaseous Oxyd of Azote or of Nitrogene*, in which he argued that a certain gaseous compound of nitrogen and oxygen was responsible for yellow fever. The man who later declared that "nature is the great experimenter" certainly came to his inquiries with the same preconceptions as his contemporaries, especially the profound faith that the new chemistry not only could but positively would bring about a revolution in the study of disease. As he admitted in his preface to the *Remarks*, "It has a long time appeared to me highly probable, that contagion was an aëriform fluid, produced occasionally, and exercising for a season its destructive effects." Inevitably, he found what he was looking for: "In the course of my experiments and inquiries, I have become satisfied my original conjecture was right; and I have to acknowledge the pneumatic philosophy has led the way to an elucidation of this hitherto dark and intricate subject."[51]

By Mitchill's time, chemists had realized that nitrogen, or azote, and oxygen combined in "four distinct proportions," distinguished by their relative degrees of oxygenation, or the amount of oxygen each compound contained. Azote with the highest degree of oxygenation formed nitric acid, followed by nitrous acid, then nitrous gas, then finally the so-called "gaseous oxyd of azote," the principal agent in the spread of yellow fever, according to Mitchill. The first indication that the gaseous oxyd of azote produced deleterious effects on animal life came from observations that Priestley provided in his compendium of chemical experiments, *Experiments and Observations*. Priestley had been performing that classical experiment of pneumatic science—putting animals into sealed containers filled with different gases to see what happens, a technique that went back to Robert Boyle in the 1660s. In doing so, Priestley noticed that the gaseous oxyd of azote, one of the "airs" that he had isolated, possessed a singular quality—animals died in it almost immediately, but flames continued to burn as normal.[52] Years later, Lavoisier noticed the same strange quality of this particular compound.[53] Almost certainly Mitchill, Priestley, and Lavoisier all referred to what modern chemists recognize as nitric oxide (NO), a simple molecule that forms a colorless, odorless gas at room temperature, proves remarkably harmful to animal life, and is not itself flammable, but "will accelerate burning of combustible materials."[54]

With the knowledge of a gas that could cause such harm to humans yet remain hidden from their detection, Mitchill proceeded to investigate the links between gaseous oxyd of azote and the incidence of yellow fever. First, he had to establish that the gaseous oxyd of azote—for which he proposed the alternate term "septon"—could be produced through natural processes, not merely artificial ones. This he did, not through experimental research, revealingly, but through a kind of deductive or analogical reasoning. On the authority of several noted chemists, including Lavoisier, Pierre Macquer, and Antoine Fourcroy, Mitchill noted that nitrous acid, a close relative of septon, derived commonly from animal putrefaction. He claimed too that nitrous acid had been detected at "the bottoms of graves where human bodies have putrefied."[55] Since nitrous acid only "barely" differed from septon in their respective degrees of oxygenation, he concluded "*a fortiori*" that natural processes must also be capable of producing septon. Mitchill marshaled evidence that linked the putrefaction of bodies with the appearances of diseases that appeared strikingly similar to yellow fever. Fourcroy mentioned that the exhumation of the bodies at the Cimetière des Innocents in Paris in 1786 resulted in a "deleterious production"—"a gas of precisely the same origin and qualities" as septon, according to Mitchill—that afflicted many Parisians.[56]

To cinch his argument, Mitchill noted that yellow fever occurred in the same places that favored the production of septon: "in large cities [where] it is generally most abundant, by reason of the greater collection, along some of their streets, sewers, wharfs, docks, &c. of those materials, which afford it, and on account of the difficulty of ventilation, in certain lanes, yards and alleys, which allows the noxious vapours to settle there."[57] In other words, cities made ideal breeding grounds for septon because they contained many sources of animal putrefaction and many enclosed spaces that could trap the poison in its gaseous state. Furthermore, Mitchill added, septon normally prevailed when the temperature hovered between 75 and 85 degrees Fahrenheit, the very same circumstances in which yellow fever prevailed.[58]

Mitchill's argument came ready-made with a public health remedy. Since septon was of an acidic quality, Mitchill simply recommended cleaning the city with water and alkaline substances, such as lime, magnesia, and potash, which would counteract the pestilential qualities of the "gaseous oxyd of azote."[59] Besides encouraging greater environmental vigilance among citizens and advocating sterner governmental regulations in regards to cleanliness, Mitchill also called for a new water supply system. "The great means of cleanliness, if properly applied, is WATER," he wrote in his *Hints toward Promoting the Health and Cleanliness of New York City* (1801). His plan called for an intricate system of aqueducts and sewers capable of carrying water from the rivers that surrounded Manhattan to the city, and then "through that medium, to convey away also those impurities, which otherwise, by floating in the air, would probably be inhaled by the lungs, and so enter into the system, and appear in a variety of shapes to baffle the skill of the physician."[60] The war against septon was to be fought with alkalines, water, and civic ingenuity.

Mitchill's doctrine of septon enjoyed wide popularity during the epidemic period. Many of his contemporaries believed that he had actually identified the chemical agent responsible for yellow fever. In the minds of a few commentators, Mitchill's demonstration obviated the need for future inquiry into the chemical nature and origins of the disease. In one of his essays, Edward Miller, a founder and editor of the *Medical Repository*, informed his readers that he would not discuss the chemical constitution of the "morbid cause" of yellow fever, as that subject had already been treated convincingly.[61] Mitchill's contemporaries, from the common country doctors to the scientific luminaries of his age, lavished him with praise. Despite his own rough treatment from Mitchill and the staff at the *Medical Repository*, Ouvière heaped praise on the rival theory. "Professor Mitchill has proved to the world what are the sources, combinations, and venomous effects, of septon," Ouvière announced in typically dramatic fashion. "The identical nature of our epidemic, in different years and places," he continued, "proclaims the truth of the discovery, and I adhere to the doctrine."[62] Priestley added his own praise to Mitchill's efforts. In a letter to Mitchill, a portion of which Mitchill reproduced in the *Medical Repository*, Priestley declared the high value of the theory—Mitchill had "made a most important discovery, which ranks with the most brilliant that this age, fertile in discoveries, can boast, that the cause of contagious fever is of an *acid* nature, and some modification of the *nitrous*, which may, properly enough, be called *septon*."[63] In the years following the publication of the *Remarks*, several inquirers came forward with analyses of the qualities and peculiarities of septon, proving that many had incorporated Mitchill's idea into their intellectual repertoires.[64]

Whatever its reception, Mitchill's theory and the methodology that produced it certainly diverged from the prescribed practices of the French chemists and, indeed, from his own epistemological ambitions. "Experiment must precede conjecture," he asserted. "Philosophers are no longer permitted to descend from generals to particulars, shaping them according to preconceived notions of their intimate relations; but are expected to proceed by a rigid examination and cautious assemblage of particulars to every general inference."[65] Yet Mitchill conducted no experiments, and his argument rested on and scattered observations drawn from published texts, and a logical deduction: *septon is a chemical compound that is very similar to another compound that occurs commonly in the decomposition of animals; the decomposition of animals and the appearance of diseases such as yellow fever have been observed to be conjoined in time and place; therefore, septon causes yellow fever.* This was a far cry from Lavoisier's insistence on laboratory precision and the primacy of experimental observation.

Despite his enthusiasm for the new system of chemistry, the severe laboratory regimen of the French reflected a view of the world and the human mind that Mitchill did not share. His approach to science remained fixed to a Christian conceptualization of nature and man, which took shape in opposition to many of the ideas associated with French philosophy. Writing years later, a contemporary and biographer recalled that when the "scepticism of France had diffused itself among all classes of the Christian world," Mitchill strengthened his Christian faith and intensified his "inquiries into the nature and designs of Providence."[66] In the *Remarks on the Gaseous Oxyd of Azote*, Mitchill explained how the theory of septon conformed to his notions of science and nature more broadly:

> The business of science is to generalize facts, to class phenomena under distinct heads, and show their dependence upon a common principle or cause. Accordingly, in the progress of human reason, polytheism has yielded to the conviction of the existence of *one God*; the intricate and seemingly opposite phenomena of matter and motion have been referred to one general law of gravitation; the puzzling and diversified appearances of electricity have been reduced to a few plain rules; the multitude of facts concerning light and colours have been in like manner arranged into scientific form; and both the *rainbow* and *telescope* bear witness to the simplicity of optics. The fluids composing our atmosphere have been analized, and the influence of these, and of many occasional combinations of other substances into gases, upon

life and health, been investigated to their principles. Contagion alone has remained a subject for doubting and guessing; a dismal somewhat, whose exact origin was unknown, and whose operation seemed capricious or unaccountable. This, I trust, will now, like other agents in creation, be found to have its laws of production, diffusion and action, which are steady and unvaried in their nature, as well as simple and easy to be comprehended.[67]

Like those of his contemporaries, Mitchill's approach to natural inquiry rested on faith in God and common sense. Nature bore witness to God's design, and that design offered itself to the inquiring human mind with its inherent faculties. Mitchill's world was simple and understandable, and it yielded truth without the severe methods of the French. Why would understanding the cause of yellow fever be any different?

To understand better what this speculative brand of chemistry offered, we would do well to examine what experimental chemistry did not offer. Despite their curious lack of prominence, two investigators composed treatises in which they described their experiments on yellow fever. Isaac Cathrall offered the first in a paper he read before the American Philosophical Society in 1800. By that time, Cathrall was a well-known figure. A practicing physician, he had served at the makeshift hospital at Bush-Hill in Philadelphia in 1793, and he later wrote a treatise about the outbreak, which he deemed to have been imported, based on the greater number of "facts."[68] Though unschooled as a chemist, Cathrall apprenticed under Dr. John Redman in Philadelphia before studying for a time in Edinburgh and then London, where he trained under the anatomically minded doctors and surgeons of the London hospitals.

It was clearly the London medical scene that had a greater influence over Cathrall. With its hospitals and anatomy theaters, including the famous ones operated by brothers William and John Hunter, not to mention its relative abundance of bodies, living and dead, London offered the best hands-on experience in the technical skills of surgery and anatomy. Upon returning to Philadelphia, Cathrall became a noted surgeon and a skillful anatomist. In his first fever treatise of 1794, he revealed his interest in the bodies of the dead. Claiming that he "spared no pains; nor . . . shrunk from scenes truly dreadful," Cathrall described the dissections that he personally performed on the corpses of yellow fever victims in systematic, yet brief and unembellished, detail. Dissection, after all, was not a thing taken lightly in the 1790s United States, where many believed that autopsy violated the sanctity of the body.[69] Cathrall's autopsies revealed an unusual sensitivity to material aspects of yellow fever, and a predilection for the type of hands-on experience that yielded palpable and quantifiable evidence.

Cathrall's education highlights the existence of a scientific-medical tradition that rivaled those taught in universities. As the closest thing to an "experimental" approach to medicine, anatomy offered a practical and profitable alternative to the philosophical and systematic approaches of university teachers.[70] Naturally Rush viewed the London scene with condescension. Acknowledging that London practitioners had produced "very useful facts," Rush tempered his praise, explaining that "few of them indeed practise medicine upon philosophical principles."[71] Rush's comment reflected a physician's prejudice toward surgeons and anatomists, who were sometimes dismissed more as tradesmen than philosophers. It reminds us too that the contagionists were more closely connected to these lay traditions, and the London milieu of medical thinkers more broadly. It is a striking feature of the contagionists—whether it was Currie and Cathrall in the States, or English commentators such as John Haygarth, an advocate for smallpox inoculation with decades of hospital experience, and Colin Chisholm, a surgeon for the British West Indian armies, and even James Tytler, the ill-fated historian whose expertise also lay in surgery—that they each practiced surgery and otherwise dealt more intimately with bodies, tissues, and liquids, and with the factual (as opposed to the theoretical) bases of medical research.

Cathrall's experiments centered on the properties of black vomit. He began haphazardly by investigating the properties and nature of the vomit. He mixed it with a variety of substances—lime water, sulphate of iron, muriated barites, nitrated silver, fixed alkalis, oxalic acid, and others. He heated the vomit over fire, boiled it in water, and froze it. He strained the liquid, separating it into its components, and then subjected each part to tests of their own. The experiments came with a certain degree of perceived risk. Writers on yellow fever since Pouppé Desportes in the mid-eighteenth century believed that the black vomit had some kind of corrosive effect.[72] The vomit did occur at the very last stage of the illness, and it almost always either preceded or succeeded the death of the patient. Thus, observers reasoned that the vomit may have caused death. Cathrall himself admitted to being fearful of the foul excrement and avoiding it for the first several years of the epidemic period. His initial experiments, on the other hand, indicated that the black vomit was a slightly acidic substance, composed mostly of water and nonreactive with his array of chemical mixtures.[73]

Having decided that black vomit did not possess the corrosive qualities imputed to it, Cathrall proceeded to test its effects on the living system. He enlisted the aid of Mr. Joseph Parker, an "active and intrepid" member of the board of health of Philadelphia. Together the two men tasted the black vomit and applied it to various places on their skins, but without any effect. Next, Cathrall and Parker placed three cats in a room and fed them a mixture of beef and black vomit. The cats, confined in such a way for sixteen days, showed no ill effects and were released in good health. Cathrall repeated the same experiment with a dog. As he described the scene to the audience at the American Philosophical Society, "A large dog was confined in a room, and, by an assistant, his jaws were forced asunder, and he was compelled to swallow an half-pint of black vomit." Aside from a bout of diarrhea, the dog escaped injury. Finally, Cathrall devised a way of testing the effects of the black vomit in its aeriform state. He evaporated fresh black vomit in a container over a medium heat while Mr. Parker "held his head over the vessel for some minutes, so as to inhale the steam of black vomit." The intrepid Mr. Parker experienced no ill effects. Cathrall himself repeated the experiment in a modified form. He enclosed himself in a small room saturated with the steam of the evaporated black vomit, where he remained for one hour. He left the room thoroughly disgusted, but otherwise unharmed.[74]

In the end, Cathrall had shown that the black vomit could not impart sickness in its fluid or gaseous state, but he did not know what it meant. Did his experiments suggest that the disease was not contagious, since the substances emitted from patients could not infect others? Or did they demonstrate that the fever could not spread through the air, because the gaseous form of the black vomit did not infect his subjects either? Perhaps the black vomit was something entirely different from the original cause, or perhaps Cathrall's adherence to contagionism prevented him from reaching conclusions.

If Cathrall's experiments had raised more questions than they answered, they also spurred further inquiry. Not long after Cathrall published his analyses of the black vomit, an anonymous author came forward with a similar treatise, which featured decidedly more definite conclusions. The experimentalist began with the goal of determining whether or not black vomit could spread yellow fever. Like Cathrall, he began by testing the black vomit in every conceivable way. He then proceeded to apply the vomit to his subjects. Again like Cathrall, the unknown investigator used cats and dogs in these studies, but with an interesting twist. Rather than forcing the animals to eat the black vomit, he inserted the substance subcutaneously. His description of the experiment offers a vivid account of scene. Under the heading "Experiment III," he recorded the following: "Having made a large incision into the back of a dog, and dissected the skin off from the cellular membrane and muscles, thereby forming a cavity, into which I poured one dram of fresh black vomit (obtained from a patient in the city hospital, who was in the last agonies which precede dissolution) and drawing the skin together, kept it in that situation by means of the dry suture; a pledget of lint was applied over this, and a bandage passed round the abdomen and over the part. The dog was confined, and prevented from irritating his back by rubbing it."[75] The dog soon recovered in perfect health.

Not content with this demonstration, the experimentalist continued with his experiments, some of a particularly cruel and macabre nature. In his fourth, he described a disturbing scene, made worse by the detached manner in which he depicted the experiment—"the jugular vein of a dog was opened, and one ounce of black vomit injected into it; he immediately shewed signs of great uneasiness, puked and purged violently, became convulsed, and expired in ten minutes in great agony." As if that were not enough, he repeated the same experiment with water and produced the same result.[76] Two dogs later, he could at least conclude that the jugular veins of animals do not kindly accept foreign liquids.

The anonymous investigator also pursued experiments on other bodily fluids of yellow fever victims. In one, he took the blood from a patient who was in the first stage of yellow fever and then inserted four drops of it into a wound on his leg. He repeated the same test several times and he drank "considerable quantities" of the blood, but without effect. Next, he enclosed quantities of saliva from a yellow fever patient in several incisions on different parts of his body. Nothing happened. He repeated the experiment with the perspiration and bile of another victim but remained in good health. In the final experiment, the urine he inserted into his wounds produced a slight inflammation, but it quickly subsided, leaving the experimenter healthy and happy.[77]

The facts, in the mind of this experimenter, led to but one "natural" conclusion—yellow fever could not be contagious. He certainly had a compelling logic to support his conclusion. Long experience with smallpox showed that certain excretions from its victims transmitted the illness, and it seemed to many that the fetid eruptions from buboes spread the plague. If yellow fever were similar to contagious diseases, as the contagionists always maintained, then it stood to reason that its victims would produce contagious bodily fluids. The experimentalist's course of experiments constituted a well-conceived, thorough, and heroic effort. Yet neither his firm conclusions nor the decidedly less certain conclusions of Cathrall appear to have had any discernible effect on the course of the yellow fever debate. Compared to the sometimes popular, sometimes controversial, always provocative works of Rush, Ouvière, and especially Mitchill, the experiments of Cathrall and the anonymous investigator went unacknowledged by their contemporaries, as far as the evidence suggests. American practitioners of science and their audiences simply did not possess deep appreciation for experimental science, at least not the type that excluded the public from participation.[78] The sharp difference between theoretical and experimental chemistry must have seemed striking to those who read the accounts of the experimentalists. One produced exalted views of nature that resolved the existence of the particles that caused yellow fever seamlessly into the cyclical and harmonious operations of all nature. The other confined enlightened inquirers in cramped laboratories and had them taste vomit and torture dogs.

The application of the new chemistry constituted a decisive weapon for the localists in the struggle to determine the cause of yellow fever. With its revelations about air, elemental structures, and chemical processes, the new chemistry supplied a compelling conceptual language with which to describe the reactions that produced yellow fever and to imagine its exact chemical structure. Localist theories suited the common-sense approach to science because they suggested that the answer to the yellow fever question lay in the design of the world, and that inquirers could grasp the answer with common mental abilities. For the contagionists, chemistry offered no substantial benefits. They believed that the material cause of yellow fever was a particle of some sort, and thus at least theoretically subject to chemical analysis, as the ill-fated experimentalists tried to show. They also admitted that the composition of the air might facilitate the spread of their contagious particles. But the contagionists' particle was still something in the air, not a part of it, and therefore largely beyond the suggestive power of pneumatic chemistry.

Through the lens of chemistry, differences between contagionists and localists come into sharper focus. At the simplest level, localists far more avidly pursued chemical inquiries. They produced ideas that linked yellow fever holistically to cycles in nature. They thought imaginatively about sublime operations and mythic powers. Contagionists did not. With their anatomical appreciation for the fabric of nature, they did not claim to know about materials they could not isolate. Currie merely speculated that contagious particles might prosper in certain situations. He did not claim knowledge about the actual chemical makeup of his contagious particle, or the chemical processes that produced it, nor did he even suggest that chemical inquiry could yield this knowledge. Perhaps like Samuel Brown, a graduate from Harvard and contagionist, he thought that "poisons, miasma or morbific effluvia are of too subtle a nature for chemical analysis."[79] In his own chemical treatise, Cathrall freely acknowledged that his experiments were inconclusive about yellow fever, nor did he ever mention which causal theory he embraced.

Though Mitchill especially benefited in the short term, his model of chemistry did not take root. In the early 1800s, chemists from Robert Hare to Benjamin Silliman replanted chemistry firmly in the laboratories where it blossomed in the nineteenth century.[80] Mitchill's romantic sentiments survived in natural history, and they flourished again with the likes of Emerson and Thoreau. In hindsight, the common-sense chemistry of Rush, Browne, Ouvière, Mitchill, and the countless others who offered their chemical observations in periodicals did not reveal as much as they hoped. But it allowed them to find what they expected to find in nature: order and intelligence. Chemistry also uncovered the terrible and frightening powers hidden in the fabric and design of the world (realizations that led to questions of their own). Disease engaged mankind in a conflict that mirrored the more elemental struggles between life and death, creation and destruction, light and darkness. In a series of poetic verses featuring a personified septon,

Mitchill rendered this unsettling truth dramatically—"You saw . . . the pec-
cant principle of death; / Grim SEPTON, arm'd with the power to intervene, /
And disconnect the animal machine . . . Within the great DISORGANIZER
lurks, / And plans, unseen, his undermining works."[81] Mitchill's lines make
for a fitting conclusion, as there always was a bit of poetry in the science of
these early republicans.

FOUR

"Let Not God Intervene"

> With reverence and resignation we contemplate the
> dispensations of Divine Providence in the alarming
> and destructive pestilence with which several of our
> cities and towns have been visited.[1]
>
> JOHN ADAMS, *State of the Union Address, December 18, 1798*

On the Fourth of July 1799, a crowd gathered in the Brick Presbyterian Church in New York City to hear a speech from Samuel Latham Mitchill. The citizens who congregated in the church's pews knew Mitchill chiefly for his scientific prowess. The charismatic professor was even gaining something of a reputation among the public for his polymathic interests and eccentricities.[2] This particular occasion, however, called on another aspect of Mitchill's intellectual repertoire. The Fourth of July celebration had become one of the most important public festivities in the early United States—a "rite of nationalism," which bound Americans around a shared cultural mythology and reinforced national mores and values.[3] In July 1799, though, the normally joyous day had taken on an unmistakably somber tone. As Americans approached the start of the fever season, most recognized that just as there was much to celebrate, there was also much to lament. Yellow fever had only recently devastated the American port cities, killing more than 2,000 in New York and over 3,500 in Philadelphia, as well as striking in Boston, Baltimore, Wilmington, Portsmouth, and Norfolk. Besides that, for reasons that Mitchill and others would discuss, the virtue of Americans seemed to be slipping, and often with very alarming results.

In this atmosphere, Mitchill chose to remind his listeners that though they might justly celebrate the anniversary of their political independence, they still depended on a great deal. Always the natural philosopher, Mitchill stressed the dominion of natural law: "As to the general influence of light, heat, and the physical elements, which compose and actuate the universe and every part of it, these citizens are as dependent on them, as any other denomination of mortal men." Then, in a much different vein, he continued, "In like manner are they dependent upon the intellectual, designing and organizing power, which gave law to the atoms of which natural beings are composed, and assigned each its sphere of action, its relations and affinities." In a disapproving gesture toward wayward contemporary philosophers, especially those among the French, he continued, "This power they ought always to acknowledge, and not affect in the wantonness or the folly of their limited intellects, to doubt of its superintending providence, or to deny, with modern epicureans, the existence of its influence. It is a sign of great weakness, and I suspect of depravity, for a people to declare themselves independent of the great governing principle in nature."[4]

Mitchill's pious admonitions to the assembly of the Brick Presbyterian Church remind us that clear-cut distinctions between science and religion did not apply to the fever investigators, especially those who embraced localism, who were almost all devout Christians as well as natural philosophers. For them, science and religion melded seamlessly in a worldview that saw in nature the very evidence of God's design. As the supreme Author, God not only created the world, he also imbued everything in it with his will and purpose. For those who embraced the notion, common sense itself bore witness to divine purpose, for God had given humans their searching minds so as to counteract the evils that beset them. It followed that the purpose of a thing constituted a key reason for that thing's existence. Aristotle enshrined purpose into his theory of causation as the final cause, the reason for which the thing was made, its *telos*. Aristotelian notions of cause and effect survived well into the modern period. Mitchill consciously channeled Aristotle and the Greeks in his Independence Day oration. His choice of the term "atoms," from the Greek word for "indivisible," signaled his debt to his Hellenic scientific forebears for providing, if only roughly, the conceptual basis of the new chemistry and its irreducible substances. His aspersions toward the "modern epicureans"—an oblique reference to the atheists and deists of his own time—showcased his alliance with the Aristotelians against the supporters of Epicurus, a contemporary and opponent of Aristotle, who believed that atoms came together randomly without any purpose.[5]

Believing in the ubiquity of divine will led to a rather unsettling conclusion—if everything had been created for a purpose, then yellow fever, too, must have a purpose, for omniscient and omnipotent God would not allow something to come into existence by mere chance. What, then, was the divine purpose of yellow fever? Why would God, an absolutely benevolent being, create such an evil? Invariably, investigators, clerics, and laypeople alike agreed that sin was responsible, but how? How exactly did God implement his will? Contemplating the *telos* of yellow fever forced investigators to navigate a labyrinth of contemporary science, philosophy, and theology. They believed in the rule of natural law and they opposed those who claimed that God directly intervened in human affairs. But they risked straying too close to the heretical doctrines of the deists in the United States and especially Europe, who believed that the evidence of design signified the existence of a creator, but who dismissed the scriptural bases of all revealed religions and denied the specific attributes of the Christian God. Eager to prove that yellow fever adhered to the rule of natural law and demonstrate its conformity with the principles of true religion, the fever investigators turned to a kind of natural theology, a type of inquiry devoted to proving the existence of the Christian God and the accuracy of the scriptures through the evidence of design.[6]

Not surprisingly, the contagionists failed to propose an effective explanation for the purpose of yellow fever. Their silence left others to grasp the irreligious implications of their arguments. Contagionists had always pointed to commerce as the source of yellow fever. They proposed limiting ship traffic in affected cities, and they advocated for strict quarantine measures that would hopefully prevent infected vessels from entering their ports. Contagionism impugned the goodness of God, for it suggested that God had allowed the yellow fever particles to move around and infect people at random. It suggested that God did not want people to engage in commerce, a notion which struck the commercially oriented early republicans as absurd. By the 1790s, most Americans, regardless of their precise political persuasions, had accepted it as common sense that some amount of commerce was necessary to ensure the well-being of the republic.[7] The localist critique of contagionism involved not only an evaluation of its scientific incorrectness but also a judgment about its unlikelihood in God's creation, evidenced by its inability to satisfy basic theological necessities.

To the contrary, the localists, led by Mitchill, Rush, and Webster, found an elegant compromise among the various theological and philosophical demands of their Christian-scientific perspectives. As they cast it, by negligently and carelessly allowing dirt and filth to accumulate, the city-dwellers violated both scriptural and common-sense prohibitions against uncleanliness, setting in motion a chain of events that, consistent with God's design, naturally produced yellow fever.[8] Their natural theological explanation also exculpated the commerce of the early United States from the charges of the contagionists. But while the localist explanation may have vindicated the economic bearing of the republic, it rekindled classic fear about the dangers of its cities and the virtues of its citizens. To remedy the situation, the localists embarked on a vigorous public health campaign, through which they stressed the duties of all Americans, as citizens of the republic and subjects of God, to abide by sanitary regulations. What they proposed was no less than a union of science with proper religion and proper republican principles. It was a powerful argument.

Since the beginning of the Judeo-Christian tradition, believers faced a singular problem—if the one true God created the world and everything in it, then why did evil exist? Why, for example, did pestilence sometimes befall the followers of God?[9] Some found some level of satisfaction in the story of the Fall—God created a world free from evil, but curious human beings (or, more precisely, one curious human woman) violated God's one command and so consigned humankind to a world of impermanence and pain. Others held with Gottfried Wilhelm Leibniz that the world as it was constituted the best of all possible worlds, and that evil was a necessary component of a good world.[10] Leaving aside such overarching justifications, most have accounted for specific instances of evil by claiming that the sins of men provoked divine wrath, and that God directly intervened in human affairs to punish the wicked. The Old Testament brims with stories of God's retributive justice, which often featured visitations of pestilence, among other penalties. In the earliest books of the Bible, for example, we read of God sending plagues to the enemies of the Israelites. In the book of Exodus, God afflicts the pharaoh and the Egyptians for their enslavement of the Hebrews, and in Samuel God sends plague upon the Philistines for stealing the ark.[11] Just as frequently, God visited pestilence on his own people for transgressing his divine will, usually because they had worshipped other gods. Much of the Old Testament retrospectively justifies the evils that befell God's people—all the plagues, famines, and wars figure as due punishments for disloyalty and irreverence.[12]

With Hippocrates, Hellenized thinkers gained a vocabulary for explaining the natural causes of diseases, but many still cleaved to notions of divine agency. Scholarship on a range of medical and scientific topics has ably shown that the supposed disjunctures between science and religion were not as complete as once thought (and the union between scientific methods and religious truths remained particularly strong in the devout colonies of British North America).[13] Ancient, medieval, and modern disease inquirers alike regularly appealed to both natural and divine explanations, sometimes in complementary, sometimes in contradictory fashion, and the results can seem bewildering to minds familiar with a clearer disunion of science and religion. Take the thoroughly Hellenized Boccaccio, writing after bubonic plague struck Florence in 1348, who claimed that the scourge came about "either because of the operations of the heavenly bodies, or because of the just wrath of God mandating punishment for our iniquitous ways."[14]

Interpretations of divine will found an especially receptive environment in the United States in the 1790s, a time of ferment in religion as much as in science. The early years of the nation witnessed the intensification of what would become known as the Second Great Awakening, a spectacular expansion of evangelical religious devotion among Americans of all sorts. Inspired by the American Revolution and its rejection of the old hierarchical, deferential social order, early republicans moved away from established churches and sought to cultivate their relationships with God through their personal experiences with his grace, the Bible, or else by selecting their own denominational affiliations in what was increasingly a competitive marketplace of religious practices and beliefs.[15] The evangelical upsurge, while never a unified movement with a fixed body of doctrine, nevertheless did center on a few essential tenets. As Mark Noll writes, "Evangelicals called people to acknowledge their sin before God, to look upon Jesus Christ (crucified—dead—resurrected) as God's means of redemption, and to exercise faith in this Redeemer as the way of reconciliation with God and orientation for life in the world."[16] Sin, punishment, and redemption lay at the heart of American religious convictions during the epidemic period.

Unsurprisingly, when yellow fever struck Philadelphia in 1793, many ascribed the pestilence to divine wrath. In a pamphlet titled *An Earnest Call*, an anonymous author urged his readers to reform their ways, as the "wrath of the Almighty seems inflamed against this City." He continued in a torrent, enumerating the crimes of the people in vague though powerful language, "His long suffering patience is at length exhausted—his mercies slighted—his gospel despised!" "At length the sword of his indignation, the sharp two edged sword of wrath is unsheathed." The unknown author even specifically set his own depiction of the cause of the yellow fever epidemic against those of the investigators: "I know there are many who attribute this awful Contagion to natural causes, and ridicule the idea of a supernatural agent: but I conceive, we may clearly trace the finger of God in our chastisement."[17]

Appeals to divine authority persisted well beyond the epidemic of 1793, even among prominent clerics and intellectuals such as Ashbel Green. A Presbyterian minister, Green later served as president of Princeton University from 1812 to 1822, where he sustained the university's reputation as a center of empirical, inductive science.[18] In *The Pastoral Letter*, the transcript of a sermon he delivered before his congregation in Philadelphia during the epidemic of 1798, Green baldly stated that God had sent yellow fever as a punishment for sins. "It has pleased a wise and holy God to lay his chastising hand more heavily on our happy city for two months past, than perhaps at any former period," Green lamented. "It would really seem as if the God of Heaven had set himself to punish the cities and towns of the United States, and was determined to inflict one stroke after another, till they were either reformed or utterly destroyed." Unfortunately, the citizens of the major port cities had not taken the lessons to heart, but had sunk even deeper into their iniquities as the epidemic period proceeded. Their depravity justified severe consequences. Drawing upon the millennialist rhetoric then common, Green intoned to what must have been a frightened audience, "It is predicted that 'in the last days perilous times shall come.' Those days it is our lot to behold."[19]

Scholars of yellow fever in the early United States certainly have not overlooked the persistence of divine causal explanations, but they have mistaken its true extent. Most simply attribute such beliefs to backwards elements in society—the people far outside of medical and scientific circles. According to Martin Pernick, faceless voices spoke about the "wrath of Deity" and its influence in the origin of yellow fever. But Pernick also dismisses such notions as the surviving relics from an older time. In the Philadelphia of 1793, he writes, "The division between medicine and theology was still young."[20] And in a different type of study altogether, Gary Nash claims that the leaders of Philadelphia's free black community, Absalom Jones and Richard Allen, viewed the plague as a visitation from God, though not as a punishment but as an opportunity to advertise blacks' civic virtues through their assistance to the fever-stricken.[21]

Single commentators mixed natural and divine explanations. In his best-selling pamphlet, the *Short Account of the Malignant Fever*, Mathew Carey alluded to the pestilence of 1793 as a providential retaliation for sins. In their newfound prosperity, many Philadelphians, Carey deemed, had sunk into luxury and extravagance. Their excesses provoked the divine wrath— "Although it were presumption to attempt to scan the decrees of heaven, yet few, I believe, will pretend to deny, that something was wanting to humble the pride of a city, which was running on in full career, to the goal of prodigality and dissipation."[22] Yet, later in the *Short Account*, Carey easily transitioned into a discussion of the natural causes of the fever. Carey also published a short pamphlet on the cause of the fever, *Observations on Dr. Rush's Enquiry into the Origin of the Late Epidemic Fever in Philadelphia*, in which he attempted to refute Rush's theory of local generation in favor of his own conviction of the fever's importation.[23]

The same curious combination of the natural and divine appears even more distinctly in the letters of Benjamin Rush. Though Rush was well known for his numerous published writings about the natural causes of yellow fever and his strenuous defense of localism, his letters to his wife, Julia, preserve an entirely different conception of the pestilence, its cause, and its meaning. In his almost daily correspondence during the epidemic of 1793, Rush repeatedly thanked God for preserving him against the plague. "I continue to enjoy good health. Help me to thank the divine Preserver of Men for it," he wrote to Julia as early as August 26. Weeks later, in early September, Rush again praised God for the deliverance from the evil, this time by quoting from Psalm 57—"'Be merciful unto me, O God! be merciful unto me, for my soul trusteth in thee, yea in the shadow of thy wings do I make my refuge, until these calamities be overpast.'" In another biblical reference, Rush compared himself to the three young men whom God preserved from the furnaces of Nebuchadnezzar for their refusal to worship a graven image of the Babylonian king—"Hereafter my name should be Shadrach, Meshach, or Abednego, for I am sure the preservation of those men from death by fire was not a greater miracle than my preservation from the infection of the prevailing disorder." Once again, on September 13, "*Alive!*" he exclaimed at the beginning of the letter, "And . . . still through divine goodness in perfect health."[24] Clearly, Rush thought that God had somehow saved him from the pestilence in the midst of which he lived and walked for months.

Indeed, according to Rush, God had also caused the disease in the first place. Though in all of his publications Rush posited natural causes for the origin of yellow fever, throughout his correspondence he very clearly and repeatedly highlighted its divine origin. In more than one instance, for example, Rush referred to the epidemic as the "judgment of God upon our city."[25] Like his contemporaries and predecessors, Rush believed that God had sent yellow fever as a punishment for sin. "What a bitter thing must *sin* be to deserve even such a punishment as a destroying pestilence," he wrote to Julia on September 30, near the peak of the epidemic's destruction. Rush even gave an indication of the types of sins that had brought about God's wrath. In one letter, he claimed, "I wish landlords would consider the wickedness of *rack* rents. They have been one of the procuring causes in my opinion of the late judgment of God upon our city."[26]

Rush's contradictory mixture of natural and divine interpretations resolved in a view of the unfolding of providence. Again and again in his correspondence he cast himself as the hero in a divine plan. His own "miraculous" survival through the epidemic, when so many of his fellow doctors died, seemed one sure sign of his role. His mother and sister's refusal to leave the city and abandon him seemed another. They had stayed to give their beloved son and brother solace enough to continue his good works and to fortify his body against fear, a predisposing cause of the illness: "My mother and sister are part of the means that providence employs to preserve me from the infection."[27] Providence operated through other agents too, and not always for the health and well-being of his subjects. God had sent yellow fever as a punishment for sin, and so in Rush's mind, the physicians with their false cures and the hawkers of quack nostrums amounted to the villains in the divine drama. Their misinformation and lies deluded the people and exacerbated the plague, but all were consistent with God's plan. "God's will is done on earth as much by pestilential contagion and ignorant physicians as it is by the songs and praises of saints and angels in heaven," he wrote.[28] And to Elias Boudinot, "Why complain of the ignorance or malice of my brethren? They are a part of the instruments of the divine displeasure against our wicked city."[29]

Rush's providential views produce more questions than answers. Did Rush really believe that God's plan unfolded during the 1793 epidemic, or did his zeal flatter Julia's piety, or perhaps his own? Did God directly intervene in human affairs to inspire or bewitch his agents? Or had God somehow preordained what was to happen, planning people and events so thoroughly that he could let them play out? And, if so, what did that suggest about human free will, and about the chains of contingent, secondary causes that natural philosophers studied? Providence was a problematic concept that spiraled down into the morass of theology. If not for subsequent developments, these outstanding questions might have gone unanswered.

James Tytler's *Treatise on the Plague and Yellow Fever* may well have incited the furor. Tytler, recall, argued that God sent contagious diseases to mankind for their sins, and that they had survived ever since, circulating among the various peoples of the world and occasionally blossoming into full-blown epidemics. *The Treatise* provided an answer to one of the glaring weaknesses of the contagionist argument—the reliance on an infinite regress, a notion implicit in the contagionist argument, which held that since disease always came from some other source through importation, it had no original cause—but the answer came only by appealing to divine intervention. In their review of *The Treatise*, the editorial staff of the *Medical Repository* praised Tytler's efforts and intentions—they even compared him to Euclid and Lavoisier, fellow compilers who had assembled and arranged the knowledge of their respective subjects—but they moved quickly to a stern rebuke of the philosopher. "Our readers will be disappointed, and, we fear, mortified, at the inferences which Mr. T. draws from the historical records." Their criticism, of course, centered on Tytler's attribution of yellow fever to God's intervention, an idea that betokened a backwards state of philosophy, they deemed. They admonished him with a Latin quote from

Horace: *Nec deus intersit, nisi dignus vindice nodus*—"let not God intervene, unless the connection is truly worthy of such an intervention."[30]

Webster's *Brief History* elicited a similar response, though for much different reasons. Webster did not directly attribute the cause of yellow fever to God, but he did promote an idea commonly known as "equivocal generation." The theory of equivocal generation held that living beings originated without a discernible cause. In the nineteenth century, it took the name "spontaneous generation" and became one of the most controversial issues of the time. It died a slow death, expiring for all intents and purposes sometime in the 1860s or 1870s with the rise of microbiology and related fields.[31] To be sure, Webster did not claim that disease matter arose spontaneously, for that would require that he thought of it as life, a notion still way off. He only claimed that the insects and mosquitoes that appeared in large numbers during epidemics in all parts of the world arose "equivocally," nurtured into existence by that occult quality of the atmosphere and explanatory catch-all—the epidemic constitution. (Ironically, Webster believed that disease produced the insects, when in fact the insects transmitted the disease.)

Despite their enthusiasm for the *Brief History*, Webster's contemporaries rebuked his advocacy of equivocal generation. In the *Medical Repository*, Mitchill and Edward Miller expressed their "surprise" at Webster's endorsement of the doctrine. According to them, the experts had "exploded" the doctrine, meaning that they had convincingly disproved it. Microscopes revealed that even plants grew from seeds, and though the editors admitted that philosophers still debated the means through which *animalcula* generated, they confidently asserted that they would one day identify it with as much accuracy as they had discovered the causes of other natural phenomena.[32] Joseph Priestley added to the censure. In a letter to Rush, later published in the *Medical Repository*, Priestley expressed his dismay that Webster "should advance opinions so wild and unphilosophical . . . and especially that he should be an advocate for what I thought to have been the long exploded doctrine of *equivocal generation*." For Priestley, a Unitarian minister, the problem was not so much evidentiary as theological. If taken to its logical extreme, the doctrine of equivocal generation devolved into atheism. For how was it, Priestley asked, that "various animals, the structure of whose bodies is as exquisite as that of man, all bearing marks of infinite wisdom, should arise spontaneously from the natural elements . . . which are void of all intelligence?" Priestley moved quickly to his powerful indictment of equivocal generation:

> If any one of these plants or animals, even the smallest and to appearance the most insignificant, could be formed without intelligence, from unconscious elements—[then] oaks, elms and cedars—horses, elephants and men, might have originally come into existence in the same way, and the whole universe have had no intelligent author. And yet Mr. Webster appears not only to be a believer in a supreme intelligent author of nature, but in revelation too. I am confounded when I reflect on such inconsistencies.[33]

The chastisements from Priestley and the editors at the *Medical Repository* provoked an indignant response from Webster. He scolded the pride and presumptuousness of his fellow investigators, whose arrogance veiled the actual limitations of their knowledge. "I think it more becoming the limited knowledge of man," he countered, "to acknowledge his ignorance, than to be positive on such doubtful subjects."[34] Webster too detected the irreligious implications of his opponents' perspectives. The rejection of equivocal generation led naturally, in his mind, to the denial of God's agency in earthly happenings and a rejection of revelation. Through the reductive logic of Webster, the matter called into question the very nature of the forces that governed the world, pitting science against religion. His position stressed the limits of enlightened inquiry:

> That matter can be endowed with laws, which shall operate uniformly and perpetually, independent of divine agency, may be possible, but appears to me unphilosophical. I can have no belief in permanency of duration in any being but God, and the operations of his power. The opinion that natural effects proceed from laws impressed on matter, without any direct exertion of divine power—and that supernatural effects are produced by the immediate agency of the Supreme Being—appears to me at least unfounded, and even unscriptural. The scripture generally ascribes every event directly to the first cause. . . . This view of the question is not only more pious, but more philosophical; for I no more comprehend the growth and expansion of the rose in my garden, than the creation of the earth, or the resurrection of Lazarus. The result of my philosophy is to resolve every event and operation in the universe into the direct exertion of omnipotence. And I cannot but think that the modern doctrine of nature and natural laws, which seems to exclude the divine agency from most of the operations in the universe, has furnished the most tenable ground occupied by the materialists.[35]

The dispute over divine intervention showcased different reactions to the critical perspectives of science and the Enlightenment, and those who used them to challenge traditional religious beliefs. French deism certainly terrified Americans, especially for its association with the French Revolution, but the danger struck closer to home as well.[36] "It is high time to chase the Deists from that ground," Rush wrote to Ashbel Green about the alarming rise in student deism at Princeton, their alma mater. The study of the divine belonged "exclusively to the Christians," he continued. "For everything good in man, and all his knowledge of God and a future state, are derived wholly from scattered and traditional rays of the successive revelations recorded in the Bible."[37] Not a strong contingent, American deists, like their European counterparts, held that a supreme being created the world, but they rejected the validity of any specific religious tradition, including God's supposed revelations in the Bible. Moderate deists at least maintained a respect for Christianity without compromising their own principles. Thomas Jefferson and Benjamin Franklin rejected the divinity of Jesus, for example, but nevertheless acknowledged that he served as an exemplar of human conduct. Radical deists ruthlessly exposed the contradictions and hypocrisies of Christianity, a religion built on lies and manipulations. In works such as Ethan Allen's *Reason the Only Oracle of Man* (1784), Thomas Paine's *Age of Reason* (1794), and Elihu Palmer's *The Principles of Nature* (1801), deists sought to substitute a religion of nature and reason for misplaced faith in the scriptures.[38]

The deist critique of Christian revelation focused particularly on the issue of God's interactions with humans. American deists specifically leveled their attacks against God's temporal inconsistency. In biblical stories, God constantly intervened in earthly happenings: he appeared to people in bushes, in dreams, in visions; he spoke to them; he caused disease and famine; he even manifested himself to his people in the flesh. Yet in the almost two thousand years after Christ, no one had experienced a miracle in which God directly interceded. Or at the very least, no such miracle had been memorialized in an authoritative religious text or accepted into the canon of religious belief. The temporal discrepancy in God's behavior undermined his perfection, Allen and Palmer argued in their works, for if God had to change, then he could not be perfect.[39] Since God must be perfect, the revelations must be wrong, they concluded. Issues about divine intervention and God's purpose were central to the deist critique of revealed Christianity.

As different as they were, both sides of the debate about divine intervention represented ways of rescuing natural inquiry from the clutches of deists. On one side of the argument, Webster denied that the autonomous operations of natural law could account for the varied phenomena of nature, such as equivocal generation. He argued that effects did not proceed from "laws impressed on matter," as though the world were merely a machine crafted by God and then left to its own devices. Rather, he suggested that God himself maintained natural laws through his active superintendence over the world (as though he so thoroughly suffused the world that he was immediately present in all of it), and that he could freely choose to contravene his normally lawful behavior. Webster's theology placed him squarely in the fold of the orthodox, Congregationalist milieu from which he came. He warned his critics that a slippery slope led to deism, with its impersonal creator, and to materialism (an idea tantamount to outright atheism). For if God had set the world in motion only to let it operate by irresistible laws, then could anyone be certain that this God was *the* God, and the revealed truths of Christianity were in fact true?

On the other side, Mitchill and Miller proved more accommodating to scientific perspectives. They claimed that nature operated according to laws that did not change, and that did not require divine superintendence. They never offered decisive word on the issue of scriptural miracles, perhaps to leave some wiggle room, and thus did not answer the question of deists' temporal inconsistency. But they indicated that God either could not or simply did not intervene to suspend his laws for any purpose, and they sought to eradicate appeals to divine intervention from the realm of legitimate inquiry.[40] "*Nec deus intersit*—let not God intervene," they declared. They argued that to admit breaches in the lawful order of the world impugned the omnipotence of God and led to atheism. For if there were no order in the universe, then perhaps there was no God.

The religious situation of the 1790s demanded a middle ground between Christianity and science. Localists were well equipped for the task—they were attuned to the design of the world, and open to common-sense perceptions of that design as truths. The outlook licensed them to move beyond the facts of the disease, to lay aside the fields of study through which facts were interpreted, and to consider the natural theology of the yellow fever. A term coined by the English clergyman and natural philosopher William Paley in his *Natural Theology; or, Evidences of the Existence and Attributes of the Deity* (1802), natural theology purported to reveal God's existence and nature through the evidence of design. Whereas deists looked to the evidence of design to rationalize the creator, American natural theologians from Rush to Mitchill reconciled design with revelation in an effort to understand God and find purpose in the created world.

Using the *Medical Repository* as the platform from which to expound right philosophy, as much about science as theology, Mitchill articulated the localists' explanation for the *telos* of yellow fever. Natural laws themselves were the means through which God fulfilled his purposes. He lectured, "In accomplishing *ends* he makes use of *means*; and these means, in the case before us, are what are termed *secondary* causes." Naturally he linked the purpose of the disease with sin: "As the *moral* conduct of man is considered instrumental in stirring up plagues, this, we conceive, may happen through a neglect of cleanliness, or by suffering pestilential filth to form and accumulate in and around their persons and habitations."[41] Mitchill stated the matter even more bluntly, "We consider it a part of the moral law, to be clean and free from defilement. The breach of this law is a vice or evil, for which the offender must expect some kind and degree of punishment." On a larger scale, the steady accretion of such sins resulted in the full-fledged epidemics that afflicted the port cities: "Noxious or pestilential vapours are the natural offspring of perspired and excreted substances undergoing corruption. . . . When, from a long-continued and excessive accumulation of such materials, and a neglect of the means of preserving public as well as private cleanliness, sickness arises, it may be properly enough considered a visitation upon the inhabitants of a town or place, for their contempt of a high moral obligation."[42] And in another work: "I consider cleanliness in our person, clothing and habitations, to be a matter of moral obligation; and the punishment which providence has wisely thought proper to inflict upon those who violate this law is sickness, not unfrequently terminating in yellow-fever, pestilence and plague."[43]

Mitchill's natural theology of yellow fever expertly navigated the various philosophical and theological pitfalls of the early national intellectuals. It upheld the rule of natural law (the laws of pestilential fermentation produced yellow fever), it acknowledged God's overarching providence (God had created yellow fever with purpose), and it redeemed God's benevolence (uncleanliness was a violation of a just command). With its emphasis on human sin, Mitchill's explanation satisfied a deeper demand of Christian

natural philosophers, for it accorded well with scriptures. Cleanliness constituted one of the fundamental imperatives of the Judeo-Christian religious
tradition. "Wash yourselves and make yourselves clean," exhorts the author
of Isaiah 1:16. Cleanliness laws figure particularly prominently in the earliest
books of the Bible, such as Deuteronomy and Leviticus, putatively written
by Moses, who enjoined his followers, in the name of God, to avoid impurities in food, in people, and in animals. Mosaic law, as it came to be known,
prescribed a comprehensive dietary regimen known as the *kashrut*, the basis
of modern kosher law; it ordered the followers of God to avoid dead bodies,
excrements, and women who were or had been menstruating, among other
things; and it outlined purification rites for those who violated the rules.

Mitchill made overt what before had been implied in the investigators'
allusions to divine punishments. More and more the theological under-
tones of the localists became explicit. And though no one quite articulated
Mitchill's natural theology as clearly as he did, many borrowed its elements.
Some called attention to the sins that caused disease. Such sins brazenly
violated scriptural law. As Rush opined, "A regard to cleanliness was enjoined upon the Jews by divine authority . . . [and] it would seem as if the
neglect of it, was necessarily connected with suffering." Though modern
scholars have debated the origins and purposes of the scriptural prohibitions
against uncleanliness, Rush left no doubt on the matter. "To prevent diseases among them, was one of the designs of their frequent ablutions, and of
many other of their ceremonial institutions."[44] Despite his public disagreement with Mitchill about the modalities of divine agency, Webster agreed
that yellow fever was brought on by transgressions of holy laws, which God
created and his prophets disseminated. "The laws of Moses were the commands of God," he declared. "It was the peculiar climate of Egypt, and the
usual prevalence of scorbutic and malignant complaints, in that country,
which occasioned all the minute injunctions of Moses, in regard to washing,
cleansing, and purifications."[45] Early republicans who had read their Bibles
should have known that God viewed uncleanliness as a serious infraction.

According to localists, common sense itself militated against uncleanliness. Believing that God had given human beings all of their emotions,
inclinations, and tendencies with purpose, localists scolded citizens for denying their instincts. Humans abhorred dirt and filth; they were repelled by
decaying matter and the odors they emitted; therefore, they must be evils.
God, they reasoned, engrafted inherent disgust onto people as a shield from

harm. Comparing the "faetor" that wafted from putrefactive material to the "rattle of the snake," Rush concluded that they were both "intended to give us notice of danger, and to remove, or fly from the filth which emits it."[46] The conclusion followed easily. Since humans naturally hated noxious materials, they would only tolerate them because of sinful behavior—vices such as greed, laziness, outright malice, or perhaps simple carelessness. Webster elaborated on the same concept: "Divine commands . . . are injunctions on man to conform to principles of moral fitness or utility, which existed *anterior to the commands*. They *unfold* to human view, and *enforce* the practice of those principles; but they do not *create* them."[47] Revealed laws only reinforced the natural order of things, making explicit what before had been implied by common sense. Webster's exegesis showcased the ultimate harmony between revelation and design.

Contemplating the natural theology of the disease compounded the problem of yellow fever. In addition to enduring a wasting disease, the inhabitants of the American port cities were dirty and sinful! Something would have to be done. From the very beginning, localists called for health laws designed to police cleanliness. Clearly localists believed that such measures would prevent yellow fever, but their recommendations, and the increasingly messianic fervor with which they expounded them, expressed other motives as well. Cleanliness imposes order; it puts things in their proper places; its goal is power and control as much as disease avoidance. When localists called for sterner rules about urban behaviors, they were not only fighting yellow fever; they were also trying to republicanize public spaces and sanctify public attitudes about filth and disease.[48] Their efforts blurred the lines between health and religion.

When yellow fever struck, the early republic was already in the midst of a revolution in personal hygiene. Washing and bathing went from "an occasional and haphazard routine" to a "regular practice of the large bulk of the people."[49] This newfound emphasis on cleanliness reflected myriad concerns: the demands of living "properly" in genteel society, a renewed emphasis on religious prohibitions against dirtiness, the profusion of disturbing new sights and smells in proto-industrial cities, and medical theories that prescribed cleanly habits as a way of maintaining health.[50] To medical authorities of the era, the dirtiness of the body constituted one of the principal predisposing causes of disease. Like intemperance, it strained the body, predisposing it to disease. Rush among others recommended "cleanliness" as the most effective measure against contracting yellow fever.[51]

Yet urbanites' deficiencies of *personal* hygiene did not address the cause of yellow fever. The miasmas of yellow fever did not exude from dirty bodies (to suggest so would have been to endorse contagionism). Rather, they emitted from masses of decaying matter that accumulated in dirty alleyways, privies, and urban sinks, among others places. Thus, while they encouraged personal hygiene as an effective prophylactic against the disease, the localists leveled their most damning judgments against the sources of social, not personal, impurities. The neglect of the body needlessly endangered the individual, but the neglect of urban filth, the true source of yellow fever, needlessly imperiled everyone. Whether it was the inconsiderate merchant who dumped putrid coffee on the wharves of Philadelphia's harbor, or the resident who carelessly disposed of his or her food, some sin accounted for each and every piece of garbage in the city, for every carcass left to putrefy. According to Charles Caldwell, a recent graduate from the University of Pennsylvania, the "evil" of yellow fever "may be said to owe its origin, in general, to that *indolence* and *inattention*, which unfortunately, constitute such predominant traits in the character of the human race."[52]

Besides transgressing holy laws, the inhabitants of the American port cities flouted republican mores.[53] In a society whose health and vitality depended on the virtues of its citizens, yellow fever and the sins that produced it posed dire threats to the fragile ideology of early national republicanism, no less than to the actual health of the nation's citizens. The problem of yellow fever raised the specter of urban multitudes run amok, the classic fear of republicanism. Fear translated into a critique of cities. In the republican imagination, cities were troubled spaces that attracted the idle and unthinking poor. On a more impersonal level, cities had basic needs. Whether their inhabitants were sinners or saints, their exhalations clogged air, their excrements filled privies, their rubbish choked alleys and contaminated waters. They required water for drinking and cleaning, plants and animals for food, and the habits of even the most assiduous workers—tanners, brewers, blacksmiths—emitted all sorts of rank excreta. Mitchill clearly recognized the unavoidable ills of city life when he implicated "the incalculable mass of animal matter afforded by the bodies of beasts, birds, fishes, and other creatures killed for the purpose of food and manufacture, and collected into cities by the industry of man."[54]

For some, like Thomas Jefferson, the troubles that arose from cities seemed reason enough to condemn them altogether. Writing to Rush in September 1800, Jefferson, who avoided public discussions of yellow fever, explained that "when great evils happen, I am in the habit of looking out for what good may arise from them as consolations to us, and Providence has in fact so established the order of things, as that most evils are the means of producing some good." He did not specifically endorse localism, but he predicted that "the yellow fever will discourage the growth of great cities in our nation, & I view great cities as pestilential to the morals, the health and the liberties of man."[55] Jefferson's interpretation of the divine purpose of yellow fever accorded well with his vision for the republic, a society, he hoped, that would be dominated by industrious yeoman farmers. It also highlights the pervasiveness and prominence of providential thinking in the early republic, an important factor underlying the localist victory. Rush's terse reply came exactly two weeks later: "I agree with you in your opinion of cities," he wrote. "I consider them in the same light that I do abscesses of the human body, viz., as reservoirs of all the impurities of a community."[56]

Abscesses or not, Rush and most other localists were unwilling to give up entirely on cities. Urbanites themselves, they put faith in public health reform and advocated stern health laws that prohibited the buildup of filth, making people responsible for their crimes. They pushed for funds and authority to employ scavengers and police to scour the streets for garbage and, hopefully, forestall dangerous accumulations of noxious materials. Urban public health reformers, such as Mitchill and Richard Bayley in New York, also called for public waterworks systems that would obviate some of the urbanites' more disgusting habits, while also serving as a source of water for the purpose of cleaning the streets, gutters, and alleyways. Localist ideas undergirded a massive effort to rejuvenate the public health apparatus in the cities of the early republic. It was a crusade fired by the religious fervor and republican sensibilities of the reformers as much as any specific notion about the natural causes of yellow fever.[57] For his own part, Webster envisioned public health reform as a return to biblical law. He called for the institution of Mosaic law in the United States: "The laws of Moses, in relation to the virtue of cleanliness, bear impressed on their front, the characters of a wisdom nothing short of divine. . . . Surely, in a matter of such essential importance to the well-being of our country, the care of legislation should be extended to supply the neglect of the moralist."[58] What the early republicans lacked in virtue, piety, and perhaps also in proper moral guidance, governors ought to make up with legislation. Webster's critique blended criticism of an immoral people with an indictment of the country's institutions, particularly its inadequate moral leadership, religious as well as civic.

The infusion of theology into the scientific discussions about the origins of yellow fever fundamentally altered the course of the debate. Increasingly, localists presented themselves to the public not merely as the purveyors of a more correct scientific perspective, but of an infinitely holier and more republican one as well. Meanwhile, they denigrated contagionism not only as an incorrect theory but as a misconceived doctrine and a negative influence on the morals of people. The inhabitants of the afflicted cities had only to adhere to localist public health measures and remove the noxious materials to free themselves from yellow fever. By doing so, they would also liberate themselves from the grievous weight of their own sins. As the epidemic period progressed, localists self-consciously transformed themselves into apostles for the union of religion with science and common sense.

The role of moral edifier came easily to Rush, a figure noted for his vigorous advocacy of moral issues. In the summer of 1799, when he offered his *Observations upon the Origin of the Malignant Bilious or Yellow Fever*, Rush tested a new means of persuasion. Writing in haste for publication before the fever season and addressing the "citizens of Philadelphia," Rush presented his *Observations* as a plea for moral rectification, lest Philadelphians suffer the same devastation as they had during the previous summer, when more than 3,500 people died. He avoided elaborate scientific discussion and instead expounded on the sinful origins of the fever. Thankfully, "to every natural evil, Heaven has provided an antidote," Rush wrote. Philadelphians could easily avoid future disaster—they simply had to obey the divine injunctions against filth and the natural effects would follow. "It is not more certain, that houses are preserved from the destructive effects of lightning by metal conductors, than that our cities might be preserved, under the usual operations of the laws of nature, from yellow fever by *cleanliness*."[59] Rush mixed scientific explanation and divine purpose in a way that simultaneously highlighted the moral complicity of Philadelphians in the rise of yellow fever and empowered them to prevent it.

But in order to abolish yellow fever permanently, Rush stressed, Philadelphians had to forsake the misconceived doctrine of contagion. Contagionism bewitched people with the illusion of their own innocence. It licensed them to continue their immoral acts. In a thinly veiled indictment of the prized contagionist critique of localism—the contention that if putrefying matter produced disease some of the time, it must produce it all of the time—Rush wrote, "The suspension of sickness from filth, no more proves it to be inoffensive, than the temporary absence of remorse for wicked actions, proves them to be innocent."[60] Like impenitent sinners, by refusing to acknowledge their sins, the contagionists persisted in wickedness.

The proselytizing spirit figured prominently in other localist writings as well. Condemning New Yorkers for their sinful negligence, Mitchill urged his readers to behave like good Christians and atone for their sins, not to blame others for the evils they had produced. For too long, he lectured, people had deflected the fault onto some exotic land or foreign people—Americans accused the West Indies of producing yellow fever, just as Europeans before them had blamed others for the plague. The righteous should confess their sins and reform their errors as part of the conduct befitting the truly penitent.[61] Like Mitchill, localists both accentuated their own superior piety and civic-mindedness and cast contagionists as an unregenerate element in society—a stubborn multitude, wedded to ancient prejudices, which simply would not listen to good sense. Following contagionism through history, Caldwell deemed that it—and especially its imputation of an infinite regress—sprang from timeless flaws of human beings. "*Pestilence* has been, at all times, treated by the world, as an illigitimate child, without an acknowledged parent," Caldwell wrote.[62] Caldwell's rhetorical strategy asked readers to view contagionists as unrepentant people who passed blame onto others but would not acknowledge their own complicity. Caldwell was also suggesting that the contagionists' infinite regress was not only a philosophical problem, one localists acknowledged from the very beginning, but a moral one as well, for it reflected the pride of humans.

Contagionists never articulated an effective counterargument to the localists' natural theological justification of disease. Their silence left inquirers to reach their own conclusions. Did contagionists agree with Tytler that God directly intervened in human affairs? And if God had initially sent disease to mankind for sins, had he somehow made a mistake in allowing it to persist? "Thus all manner of mischief sallied forth from Pandora's box; and one of these fables is about as worthy of credit as the other," the writers at the *Medical Repository* scornfully remarked of Tytler's idea.[63]

Contagionism seemingly impugned the goodness of God. Contagious particles moved around randomly, by the whims of travelers, traders, and settlers. Anyone at any time could contract contagion and then unknowingly transfer it to another place or people, even a just people. Why would God allow such an entity to exist? Contagionism suggested that God explicitly wished to discourage the intermixing of people and goods, just as the contagionists' quarantine regulations implicitly did.

And yet it seemed so obvious to mercantile early republicans that God favored commerce, whose goodness seemed evident from common-sense reflection on the design of the world. Had not God created man with the desire to engage in commerce? Experience too showed that trade redounded to the benefit of society as a whole, since it enlarged wealth, facilitated peaceful interactions among people, and spread the blessings of civilization to the benighted.[64] Speaking for the trade-oriented citizens of the fever-stricken cities, localists vigorously defended commerce. As Rush wrote: "I consider commerce in a much higher light when I recommend the study of it in republican seminaries. I view it as the best security against the influence of hereditary monopolies of land, and, therefore, the surest protection against aristocracy. I consider its effects as next to those of religion in humanizing mankind, and lastly, I view it as the means of uniting the different nations of the world together by the ties of mutual wants and obligations." Rush specifically defended commerce against the charges of contagionists: "Commerce can be no more endangered than Religion, by the publication of philosophical truth," he stated in *An Account of the Bilious Remitting Yellow Fever*.[65] Mitchill too consistently advocated for the expansion of commerce—he cofounded the Society for the Promotion of Agriculture, Manufactures, and the Useful Arts; as representative for New York beginning in 1801, he served on the Committee on Commerce and Manufactures, where he worked to remove quarantine regulations; and he later became a leading spokesperson for the Erie Canal.[66]

The contagionists' implicit critique of commerce did not help their effort to determine the cause of yellow fever, but as always their understated arguments and implications rested on deeper truths. The trade boom with the West Indies in the 1790s, and in particular the enormous riches American merchants reaped from reexporting Caribbean cash crops, piqued apprehensions about dealing with the region, whose constitutive relationship with slavery troubled many. And indeed, slavery, commerce, and yellow fever were intimately related, but in ways that early republicans may have intuited but did not fully understand. Both the yellow fever virus and its insect vector, the *Aedes aegypti* mosquito, came from Africa; they migrated to the New World in the bodies of slaves and in the ships of slavers. As for the spate of yellow fever epidemics in the 1790s and early 1800s, the slave rebellion in Saint-Domingue had brought them on by prompting massive infusions of nonimmune white soldiers—food for hungry mosquitoes and hosts for the virus. The commerce in coffee and sugar, made with the sweat and blood of slaves, brought yellow fever to American ports. That the unsettling implications of the West Indies failed to sway public opinion speaks volumes about the commercial bearing of the republic. To the contrary, localists may not have fully dispelled concerns associated with the West Indies trade and the slaves who made it possible, but their reaffirmations of the healthfulness of commerce jibed with a populace moving forward in a bustling, mercantile world.[67]

The episode involving natural theology and yellow fever reinforces how far contagionists differed from localists. In a society where people sought avidly for God's will, where God's judgment seemed to hang over everyone and everything, where many believed that the end of days was at hand, the contagionists failed to justify the theology of a disease that so clearly came as a punishment for sin. Tellingly, they never even tried. Here as in other areas of inquiry, their reticence betrayed deeper commitments to scientific inquiry. Committed to plain facts, averse to theories, contagionists deemed that God and his will fell outside the boundaries of scientific study. Like Webster's epidemic constitution and Mitchill's septon, God's purpose was not a material object they could touch and feel like the patients they attended to during their medical apprenticeships or the corpses they handled in London hospitals.

For the localists, such clear-cut distinctions among fields of study, between science and religion, never existed. If natural laws punished bad behavior, then scientific theories that explained those laws, the religious texts that prohibited those bad behaviors, and the political laws and social norms that imposed punishments for those transgressions were all linked together in a very intimate way.[68] These early republicans did not mix science, religion, and politics; their points of view suggest that these domains of thought were parts of an inseparable whole. The range of subjects and fields that fell under the umbrella of science were broader than we might think, and what united them together was common sense—both the common-sense faculty that the human mind possessed, and the common sense they saw in the world, as truths that seemed obvious from reflection on the nature of God's world. The localists' entire perspective was patently teleological, and indeed circular. Common sense told them that they could understand the design of the world, and the apparent design of the world told them that they could trust common sense.

The localists' natural theology rationalized yellow fever, but its effects were more ambiguous than it might seem. Their contemplations of the providential meaning of yellow fever forced localists to reckon with the sinfulness of the republic's citizens, as well as the influence of its commercial cities. As republican theorists had long suggested, cities were dens of vice—"abscesses" to use Rush's phrase—inhabited by sinners who lacked the public virtue demanded of republican life. Their foul habits produced yellow fever. Like the historical study of the disease, natural theological inquiry challenged the exceptionalism of the American republic. But if history suggested that the Americans were hopelessly caught up in the sweep of history, natural theology offered a solution. The inhabitants of cities could avoid yellow fever, but they had to open themselves to a religious and political, no less than a scientific, conversion. As the United States entered a new century, and as the society's leaders struggled to set the republic on a virtuous, sustainable path, it was all too clear that there was still much work to be done.

FIVE

"In Politics As Well As Medicine"; or, The Arrogance of the Enlightened

> Nor grieve that slander, with malicious rage,
> Fierce war against thy glory wage:
> Beneath a grim, ferocious sway,
> Palmyra's ruins gloom'd the desert way;
> But wide o'er earth their fame forever flies;
> Amid their venerable shade,
> Oft hath the musing pilgrim stray'd
> There, oft, the Genius of the waste,
> Wanders, & thinks upon the past;
> Counts o'er the mouldering piles, & sighs
>
> Nor yet, tho' factions' busy imps defame,
> And load with every word of shame,
> Think that the offspring of thy mind,
> Was e'er for mischief's evil designed[1]

ELIHU HUBBARD SMITH, *1795*

When Benjamin Rush resigned from the College of Physicians of Philadelphia in 1793, an institution he helped form only a few years before, he cited the "persecutions" he suffered from the fellows of the college as justification. "Besides combating with the yellow fever," he fulminated to his wife, Julia Rush, on September 13, "I have been obliged to contend with the prejudices, fears, and falsehoods of several of my brethren, all of which retard the progress of truth and daily cost our city many lives."[2] Days later, again to Julia, "They have confederated against me in the most cruel manner and are propagating calumnies against me in every part of the city." He continued,

"If I outlive the present calamity, I know not when I shall be safe from their persecutions. Never did I before witness such a mass of ignorance and wickedness as our profession has exhibited in the course of the present calamity. I almost wish to renounce the name of physician."[3] The "calumnies" directed against Rush focused principally on his "heroic" treatment—the bloodlettings and purges that left his patients dazed and depleted.[4] By November 5, when Rush tendered his letter of resignation to John Morgan, the president of the College of Physicians, the epidemic had all but ended; he might have expected the abuses from his colleagues to have ended too.

Yet, to Rush's chagrin, the persecutions continued. As he saw it, the contagionists had conspired together and were attempting to commandeer the yellow fever discourse not only by disparaging their opponents but by disseminating falsehoods that deceived the people. By 1797, the contagionist "faction," as he liked to call it, had grown so bold in its intrusions—and so successful in having its quarantine restrictions enacted—that Rush and his localist friends founded the Academy of Medicine of Philadelphia, an institution meant to rival the contagionist-controlled College of Physicians. The Academy of Medicine would liberate both physicians and the public from the tyranny of the contagionists, thereby rescuing unprejudiced medical inquiry. Rush wrote to Noah Webster, explaining the meaning and importance of the Academy: "All those physicians who believe in the domestic origin of yellow fever . . . have lately formed themselves into a medical society for the purpose of promoting medical science untrammeled by the systems of medicine which now govern the greatest part of the Physicians in our city."[5] For the rest of the epidemic period, both the College and the Academy published opposite explanations for the cause of yellow fever.[6]

Rush's feuds and schisms illustrate how the intellectual community stimulated by the yellow fever crisis broke down into rival camps. Almost invariably investigators accused each other not merely of being wrong but of representing private interests and intentionally distorting the truth. Proponents of both theories decried the opposition as a "faction," a dreaded association of self-interested individuals. The investigators' preoccupations with faction mirrored a well-known strain of thought in early republican political discourse that also centered on the intrigues of secret societies. This "paranoid style," as Richard Hofstadter called it, has appeared regularly in American political thought, but it thrived during the revolutionary and early republican eras.[7] In the early republican mind, factions lurked everywhere—in the British government and its growing bureaucracy; in France during its struggle to remake society; and among the inhabitants of the United States, in their newspapers and societies, and in their networks of correspondence.

The similarity in the tones of the discourses about yellow fever and politics indicates a deeper, structural bond and promises to reveal much about the nature of public discourse. Talk of politics and science aroused conspiratorial suspicions because they occupied the same spaces in the early republican public sphere, and they were rooted in the same material organization. In both, disputants attempted to sway public opinion by issuing pamphlets, publishing in newspapers, and giving speeches. They also formed private associations and established networks of private correspondence through which they exchanged ideas and insights, and crafted strategies. Rush's network congealed around the Academy of Medicine but spanned the entire United States, knitting together investigators such as Noah Webster in New England, Samuel Latham Mitchill in New York, and David Ramsay in South Carolina. William Currie's network ranged further yet, as it included Americans such as Isaac Cathrall as well as English physicians with "interests" in the West Indies such as John Haygarth and Colin Chisholm. Those who glimpsed these networks, who could see that secret talk generated public declarations, concluded that the individuals who made them constituted factions.[8] Perception of factions in politics and yellow fever proved mutually self-sustaining: the evident truth that factions existed in politics lent credence to the notion that they might also exist in science; the existence of factions in science reinforced that they existed in politics.

The conspiratorial tone of the yellow fever debate shows cracks and fissures in the ideals of discourse in the early republican public sphere. When investigators first offered their opinions about the cause of yellow fever, they did so believing in an ideal of rational and transparent public discourse. In the knowledge culture of the early republic, knowledge-makers submitted facts and explanations to the public for its consideration, and the public itself decided the matter. The integrity of the knowledge culture rested on the trust that when one philosopher submitted facts and explanations to the public consideration, he did so faithfully and disinterestedly. That is partially why clubbish collegiality and sociability were so highly valued in the knowledge culture of the early republic, and why individuals took such pains to emphasize their good-natured and unassuming commitment to facts and their freedom from theories.[9] But the reality of public debate in the faction-beset 1790s destroyed faith in the ideal. Tales of conspiracies and factions reflected anxieties about the transparency of public discourse.[10]

Localists turned their critiques inward, against the structures of public discourse and those who conducted it. For most, watching the yellow fever debate splinter into factions highlighted the fallibility of humans. They concluded, as James Madison did in his consideration of human political tendencies, that the tendency to form factions was rooted in human behavior. "The latent causes of faction are thus sown in the nature of man," Madison wrote in the Federalist 10.[11] But investigators could not dilute factions in a large republic, as Madison's remedy promised. Instead, the localists, Mitchill and Webster especially, directed their anger against the free press, and what they deemed an overabundance of liberties in general. They called for top-down control over medical and scientific discourses. This motive was behind Rush's formation of the Academy of Medicine, and behind the investigators' growing insistence that their opponents' perspectives were not only wrong but invalid, and that the question could be decided by qualified people.[12] There was much that was ironic about the investigators' conspiratorial imaginations: convinced that their opponents had conspired against them, they formed factions enshrouded in the secrecy and dressed with the autocratic authority over scientific matters that they reviled.

But tellingly, those who threw accusations never saw themselves as the makers of factions. In their own minds, they remained principled and honest seekers of truth, totally beyond reproach. This ingenuous self-confidence reflected a deeper faith in Enlightenment ideas that undergirded talk of science, medicine, and politics. Medicine and politics shared another similarity—both aspired to the status of sciences; both were thought to be perfectible, but only when humans had thoroughly embraced reason. Again, Madison on politics: "As long as the reason of man continues fallible, and he is at liberty to exercise it, different opinions will be formed."[13] Confidence morphed into hubris. Since there were right answers, and since everyone also *knew* that they had discovered those right answers based on their own superior evidence and reasoning, they concluded that their adversaries were not only wrong but badly, even dangerously, deluded, deceived, or ill intentioned. This was all the more true in the context of the yellow fever debate, especially among the most bullish localists, who had all along asserted that their theory reflected common sense and conformed to the way that God designed the material world and invested it with purpose. William Chalwill, a localist, encapsulated the position nicely: "It is an unfortunate truth, that the generality of mankind are blind to conviction, wherever truth and interest oppose each other," Chalwill began. "In yellow fever this is particularly the case. Notwithstanding the numerous, decided proofs, brought forward by men and literary bodies, of the first eminence, of the domestic origin of this disease, yet there are not wanting many who believe, it owes its origin to importation."[14] Here was the arrogance of the enlightened, and the intellectual life of the early republic sagged under its weight.[15]

The fear of factions, cabals, and conspiracies loomed large in the minds of Americans of the revolutionary and early national periods. It certainly colored colonists' perceptions of British imperial intentions in the troubled 1760s and 1770s. When the British imposed new taxes on their North American colonists, Americans concluded that would-be oligarchs hidden within the British government and its sprawling, unwieldy bureaucracy sought to deprive them of their rights in order to enhance their own power. Legislative acts only signaled the beginning of a much darker and more intricate plot to subvert colonial society, and to reduce the colonists to a state of virtual slavery.[16] The fear of factions also gave rise to the form of national government outlined in the Constitution. Its principal author, Madison, constructed the Constitution with the intent of limiting the power of faction, which he defined as "a number of citizens, whether amounting to a majority or a minority of the whole, who are united and actuated by some common impulse of passion, or of interest, adverse to the rights of other citizens, or to the permanent and aggregate interests of the community." Not only did factions promote unrepublican values, they could also commandeer governments and pervert even well-conceived constitutions, as the corruption of the English government illustrated. In the Federalist 10, the most celebrated of the federalist arguments, Madison contended, "AMONG the numerous advantages promised by a well-constructed Union, none deserves to be more accurately developed than its tendency to break and control the violence of faction."[17]

Sharpened by the events of the preceding decades, the conspiratorial imaginations of the early republicans flourished in the tumultuous 1790s. After the ratification of the Constitution, tensions between the nascent political groups quickly fueled suspicions about the true interests of such "factions," as their opponents deemed them. The rise of partisan newspaper conflicts and the proliferation of Democratic-Republican clubs added to the hysteria.[18] Many believed that such clubs spread discord, as in France, where the Bavarian Illuminati supposedly engineered the undoing of the revolution. From Pennsylvania, Ebenezer Hazard wrote of the Democratic clubs, "It is astonishing to see that they cannot see the impropriety of their meeting for the express purpose of influencing measures of government and forming themselves into *Societies*." He warned against their consequences, "They must be carefully watched & their machinations guarded against, or we shall yet see such scenes as have been acted in France."[19] Finally, in the South, the slave revolt in Saint-Domingue in 1791 and the Haitian Revolution provoked fears of slave revolts and treacherous mass murder. The actual plot led by Gabriel Prosser in Virginia in 1800 only confirmed such suspicions and honed slaveholders' vigilance against rebellion.[20]

In the chaotic world of the late eighteenth century, even natural philosophers fell victim to the persecutions of faction. The coming of Joseph Priestley to the United States in 1794 provided American onlookers with a dramatic example of the reality and effects of public persecution (although Priestley did not face the persecution of a faction, but of a mob who believed that he belonged to a faction!). In 1791, a riotous crowd invaded Priestley's home in Birmingham, England, and torched the house, ruining Priestley's laboratory equipment and library, and destroying valuable manuscripts in the process. Priestley, a practicing chemist, had not attracted this persecution for his scientific views, as unpopular as they were.[21] The public's ire centered on Priestley's political and religious opinions—his outspoken support of the French Revolution during its early stages, publicized in a widely circulated reply to Edmund Burke's well-known denunciation of the revolution, as well as his prominence as a Unitarian Dissenter and theologian, who wrote on subjects that contradicted Anglican orthodoxy.[22]

Shaken but determined to remain in England, Priestley moved to Clapton, outside of London, only to find that the persecutions—at the license and behest of the Court, according to Priestley—continued.[23] For a time, he contemplated moving his family to France, where he had many friends, but ultimately wound up traveling to Northumberland, Pennsylvania, where the appeal of plentiful land and the prospects of starting a Unitarian congregation won him over.[24] Besides, by 1794, when Priestley made the trip across the Atlantic, the Reign of Terror had descended and France had become a very dangerous place. Long before his move, Priestley saw signs of the coming disorder. In a letter of June 2, 1792, addressed to Antoine-Laurent Lavoisier, with whom he maintained amiable relations despite their noted scientific disagreements, Priestley noted ominously, "I shall be glad to take refuge in your country, the liberties of which I hope will be established notwithstanding the present combination against you."[25] Months later, these "combinations," let loose under the Terror, engineered the trial and execution of the chemist. Despite his well-meaning enthusiasm for the principles of the revolution, Lavoisier's years-long partnership in the Farm General, an institution that gathered taxes for the crown as well as the revolutionary governments, aroused suspicions about the fervor of its leaders. When rumors surfaced alleging that the Farm General had defrauded taxpayers and the government, Lavoisier and many others were brought before a revolutionary tribunal. Like many others at the time, he was convicted and then guillotined for crimes that amounted to conspiracy against the French people.[26] At the same moment, Priestley was sailing for the United States on a ship called *Samson*.

Given the imagined pervasiveness of factions, we should not be surprised to find that leading investigators—Rush, Mitchill, Webster, and Currie, all of them intimately involved in public life—believed themselves to be victims of factional designs. Already in the early 1790s, Noah Webster discovered "faction" hounding him for his opinions regarding the French Revolution. Barely thirty years old at the time of the storming of the Bastille, and still with his youthful exuberance and faith in human goodness intact, Webster began as an ardent supporter of the revolution and its promise of transforming the Catholic monarchy into a Protestant republic. But his active advocacy of the revolution and its goals earned Webster the mistrust of his contemporaries, at least as he saw it. Writing to President George Washington in 1790, Webster complained that "the prejudices of many men are against me" and noted that "I have written much more than any other man of my age in favor of the Revolution and my country, and at times my opinions have been unpopular."[27]

Like many other Americans, Webster attributed the downfall of the revolution to the villainy of faction. In a reactionary pamphlet, Webster deplored the tyranny of the Jacobins, the usurpers of the revolution, who sank France into terror and bloodshed. Always the historian, Webster likened the

Jacobins to the *decemviri* of fifth-century Rome—the ten magistrates granted near-dictatorial powers to draw up the Twelve Tables, the famous Roman law code, but who refused to relinquish their appointments, committed crimes against the Roman people, and were violently expelled from office. The Jacobins' predilections for deism boded ill for the virtue of the French republic and its future prosperity. Webster, therefore, used the latter parts of his pamphlet to urge his American audience to guard against "*faction*, that enemy of government and freedom."[28] By the late 1790s, Webster had allied himself more closely with the New English Congregationalist milieu that mounted the fiercest counterattack against the French Revolution.[29]

Webster's most heated confrontations with faction evolved from his participation in the vituperative political newspaper wars in New York in the 1790s. As the editor of the federalist-leaning *Minerva*, Webster feuded with the leading democratic-republican newspapers, such as the *New York Journal*, which accused him of being a sloppy newspaperman, an elitist—an "utter enemy of the rights and privileges of the people"—as well as a partisan: a mere "scribbler of a British faction." In his May 2, 1796, issue of the *Minerva*, Webster defended himself in a lengthy passage. "During the time which I have conducted the publication of this paper, the public mind has been much agitated with party spirit." As for himself, Webster averred, "In point of *facts*, my invariable rule has been to state them as I find them, and according to the best evidence obtained, without regard to any party."[30]

Webster also detected the workings of faction in the yellow fever debate. Initially receptive to contagionism and its compelling evidence of importation, Webster gradually realized, to his shock and dismay, that the doctrine rested on lies. The conversion for Webster came sometime in the fall of 1797, when he was preparing his responses to William Currie's very public pronouncements in favor of contagionism (Currie, recall, had published several letters in Benjamin Wynkoop's Philadelphia newspaper, which were then reprinted in New York newspapers). Webster focused particularly on one episode that showed the duplicity of the importationists like no other. In Currie's second letter to Mr. Wynkoop, he claimed that the yellow fever of 1793 originated from Boullam, in Africa, and then traveled across the Atlantic to the West Indies and United States. Currie derived his argument from a 1795 essay by Colin Chisholm, a doctor who had served with the British armies in Grenada during the wars of the French Revolution and a contributor to the trans-Atlantic contagionist network that included John Haygarth.

Chisholm's treatise reconstructed the circuitous path of the *Hanky* as it made its way from England to Africa and then to Grenada, bringing yellow fever along. However, since Chisholm had remained the whole time in Grenada, he based his narrative almost entirely on the testimony of J. Paiba, a passenger on the *Hanky*.[31]

According to Chisholm's rendition of Paiba's story, the *Hanky* and one other ship sailed from England at the beginning of April 1792. Both had been chartered by the Sierra Leone Company and were destined for "Boullam," the island now known as Bolama, off the coast of Guinea-Bissau. Paiba and his fellow "adventurers"—the same term that Thomas Clarkson, one of the founders of the Sierra Leone Company, used for the settlers— intended to settle Boullam as a colony for the ex-slaves rescued from their American enslavers during the American Revolution.[32] Chisholm noted that the crew of the ship were inflamed with "the fanatic enthusiasm for the Abolition of the Slave Trade," as well as the "delusive prospect of wealth" supposed to be gotten from the cultivation of cotton.[33] Once they arrived near Boullam, unfavorable circumstances forced the passengers to stay on board the *Hanky*. Boullam, they realized, was "destitute of fresh water," and the wells they dug on the shore produced only "brackish" and "unwholesome" water. Moreover, the colonizers met with resistance from some of the native inhabitants—"the negroes of this part of Africa are ferocious in an extraordinary degree," Chisholm claimed, "and are even said to be cannibals."[34]

Having remained onboard the *Hanky* in crowded, unclean conditions, the passengers contracted a disease, which promptly spread to the nearly two hundred people onboard. Pressed by these circumstances, the *Hanky* sailed to the nearby Portuguese settlement of Bissau for supplies. Overwhelmed with the sick, the ship returned briefly to Boullam, but then headed to St. Jago, the largest of the Cape Verde Islands, now known as Santiago. At St. Jago, the ship of colonizers met with two British warships, the *Charon* and the *Scorpion*, whose benevolent captains sent four men to the *Hanky* so as to help navigate the ship to the West Indies, "a voyage to England being impractical in their wretched state." Only three days out of port, the new passengers took ill; by this point, too, Chisholm claimed that three-fourths of the crew had been "carried off." The *Hanky* limped into Grenada on February 19, 1793. "From this period," Chisholm continued, "we are to date the commencement of a disease before, I believe, unknown in this country, and certainly unequalled in its destructive nature." Thereafter, the new disease, yellow fever, spread rapidly through the West Indian islands, making its way to Philadelphia in August of 1793.[35]

In his second letter of October 28, Webster gently chastised Currie for embracing the story of the *Hanky* and importation of the fever from Boullam. Webster probed the ambiguities of the story. "How and by what means did the disease at Boulam originate among the people?" he asked. It certainly could not have come from contagion, Webster opined, for Boullam was "an uninhabited island, and has no marsh grounds to vitiate the air."[36] If the sickness did not come from the environment or from a specific contagion, then Webster could only reach one conclusion—"the people themselves, to use a vulgar, but expressive phrase, *bred the plague* among themselves." But, as Webster well knew, human effluvia could only spread disease within a confined radius, as in the camp, jail, and hospital fevers.[37] Chisholm's story, on the other hand, expressly stated that the contagious particles had originated on Boullam, whence the crew of the *Hanky* carried them elsewhere. Webster's early critique of Currie, then, struck only at the major weakness of the contagionist perspective—the inability to specify an origin of the specific contagion.

Sometime before November 15, when he published his ninth letter in the *Commercial Advertiser*, Webster found a new source of information that implicated the integrity of Chisholm's account. By a stroke of luck, Webster located Mr. Paiba, Chisholm's main informant, who had taken up residence in New York City with his wife. Webster promptly arranged interviews with the Paibas, which he conducted with "a medical friend, Dr. Smith," who "made notes of the facts related."[38] Having read Chisholm's narrative of events, Mr. Paiba offered a significantly different depiction of the voyage. In his ninth letter, Webster enumerated several individual points of disagreement. According to Paiba, the expedition to Boullam, though undertaken by abolitionists, did not take place under the aegis of the Sierra Leone Company. Mr. Paiba himself, it turns out, had been one of the twelve directors of the *independent* voyage to Boullam. Contrary to Chisholm's account, the water at Boullam was not brackish, but of a high quality. Nor did the passengers stay onboard the *Hanky* for fear of the African inhabitants; rather, they settled comfortably on the land—as Webster wrote, they "cleared several acres of land, and built a block house, with four gates, and planted artillery to defend them."[39] Finally and most damningly, the *Hanky*, though troubled by illness, did not arrive in Grenada on February 19, 1793, but remained at sea until March 18 or 19, well after yellow fever had made its appearance on the island.[40] The *Hanky*, it seems, could not possibly have sparked the yellow fever pandemic of 1793.

Why did Chisholm alter significant elements of Paiba's original testimony about the voyage of the *Hanky*? As Webster saw it, Chisholm's fabrications took part in a carefully concerted scheme to paint the coast of Africa as a dangerous and unhealthy place, and thereby discourage colonization efforts that threatened the vested interests of West Indian planters, who made fortunes on slaves and the lucrative crops they produced. In their minds, successful colonization would benefit the rising tide of abolitionism in Great Britain. Planters could also do without the competition from the freemen growing cotton or any other profitable commodity on the coast of Africa. As a result, the planters and their friends in England, including Chisholm apparently, mustered the full force of their media arsenal in order to cast the Boullam enterprise in a negative light. "I will only observe in general," Webster proclaimed, "that every species of falsehood and misrepresentation, was resorted to by the enemies of the Boullam enterprise to discourage the prosecution of the settlement, which might result in putting a stop to the nefarious slave trade, and in establishing the culture of cotton by free hands."[41] By associating the "mischief" of the Boullam group with the Sierra Leone Company, the proslavery interests believed that they could effectively discredit the much larger and better organized group. For although the first settlement of the Sierra Leone Company—a place called Granville Town, in honor of one of its antislavery creators, Granville Sharp—had quickly fallen apart, the company was busily raising new settlements in Sierra Leone. If the company could be depicted as a disseminator of epidemics and bane to the welfare of the people, however, these new efforts might lose their support.[42]

The story of the *Hanky*, Paiba, and Chisholm, as well as others like it (so he claims), convinced Webster of the extreme duplicity of the contagionists and led to his final, unequivocal rejection of contagionism. In the *Brief History of Epidemic and Pestilential Diseases*, a work he imagined as an extension of his earlier letters to Currie, Webster recounted his reasons for abandoning the "ill-founded" doctrine. "I found repeatedly that the reports of persons taken ill, in consequence with vessels from the West-Indies, or with diseased seamen, infected cotton or clothing, or the like causes, were mere idle tales, raised by the ignorant or interested, and wholly unsupported by evidence."[43]

Webster might have been willing to dismiss Chisholm's dishonesty as an instance of human weakness had it not been for the forum in which it took place, the high stakes of the contest, and the dawning realization that Chisholm's alleged impropriety was all too common in the factious early republic. Yellow fever was a public debate. Newspapers and pamphlets disseminated ideas about the yellow fever to concerned readers; and the public places in the early republican cities—their piers, taverns, coffeehouses, markets, coaches—buzzed with talk of the disease. As E. H. Smith wrote in 1795, "Wherever you go, the Fever is the invariable & unceasing topic of conversation. When two persons meet, the Fever is the subject of the first inquiries. . . . In one shape, or other, the fever is constantly brought into view."[44] Yellow fever, and the fate of the debate it engendered, also had definite consequences for a wide range of people. Because of its centrality to pressing issues, from slavery to commerce, the task of ascertaining the cause of the disease could easily be misused to secure private gain. Therein lay the problem. The political wars of the 1790s taught early republicans that public pronouncements, especially those couched in terms of disinterested public charity, often betrayed private interests. Webster clearly believed that the same lesson applied to yellow fever. To him, Chisholm's public advocacy of contagionism and his conflicting private interest in the slave trade stunk of malfeasance, and the logic of the times made it seem true.

Contagionists recognized the same possibilities. They did not fail to notice that localism exculpated the commercial reputations of afflicted cities and that localists such as Rush and Mitchill advocated for the commercial interests of Philadelphia and New York. Never as brazenly accusatory as the localists, the contagionists William Currie and Isaac Cathrall did suggest that perhaps localists truly rejected contagionism because it led to quarantine procedures that derailed ship traffic in American cities and otherwise persuaded foreign ports to deal more cautiously with American vessels. Modern scholars too have found elements of self-interest in the localists' strident advocacy of localism and their seemingly irrational rejection of contagionism. For why else would they deny the well-founded evidence in favor of contagionism?[45]

Mutual distrust reflected the fact that discourses about politics and science shared structural similarities and took place in the same public spaces.

In both, discursive communities—"factions"—formed through networks of personal correspondence and private associations, and they communicated through newspapers, pamphlet wars, and even public oratory. Factions sometimes could transmit secret information, and they often spanned great distances. From the West Indies and England, Chisholm maintained connections with prominent thinkers such John Haygarth, a man who more than once insinuated himself into the American fever discussions, and his *Essay* reached audiences all over the Atlantic rim. Those who saw his work knew that they were only looking at the tip of the iceberg. They knew that the pamphlet, like any other public offering, was not only built on the evidence (the facts!) and reasonings that were presented in it; it also rested on ideas acquired through the nuances of personal experience, through casual conversations and learned exchanges, through dockside gossip and the rumors that drifted through bustling seaport cites, and of course through the deliberate, self-interested machinations of factions. Discerning consumers who could make out, if only vaguely, the hazy and circuitous route through which that knowledge traveled fretted about what it meant. They worried about what might have been said behind closed doors, about what designs lay behind scientific and medical ideas.

When Webster offered his criticisms in the *Brief History*, he did not just turn against Chisholm, Currie, and the contagionist faction, he also took aim at the structures of public discourse in the early republic. He singled out the press. Citing common anxieties about the place of newspapers in the republic, Webster opined that they "may be instruments of great good or extensive evil." He continued more directly, "It is obvious that the falsehoods and calumnies propagated by means of public papers have been the direct and principal means of all the civil dissensions which distract this country."[46] This was a bold statement. As Webster cast it, "civil dissensions" were not ultimately rooted in the canards, calumnies, and conspiracies against which he railed so fiercely—these were tragically and inevitably the occasional results of flawed and Fallen humans. Rather, Webster blamed the "public papers," no less the ones that disseminated Currie's medical lies than the ones that distributed Jefferson's political heresies. The conclusion clearly seemed a logical one—since the discourse about yellow fever was communicated through the same channels as political discourse, then it was only natural for Webster to conclude that yellow fever would propagate the same lies as the political discussions.

Faction undermined Webster's faith in people and the press, but it also sullied his opinions of the political revolution he had fought so hard to achieve. Webster reasoned that the distortions that spread through the free press represented a dangerous excess of liberty. In a letter to Benjamin Rush of December 15, 1800, Webster voiced his concerns about the structural foundations of republican life:

> As to mankind, I believe the mass of them to be *copax rationis*.[47] They are ignorant or, what is worse, governed by prejudices and authority—and the authority of men who flatter them instead of boldly telling them the truth. It is so in politics as well as in medicine. It would be better for the people— they would be more free and more happy—if all were deprived of the right of suffrage until they are 45 years of age and if no man was eligible to an important office until he is 50, that is, if all the powers of government were vested in our old men who have lost their ambition chiefly and have learned wisdom by experience; but to tell the people this would be treason. We have grown so wise of late years as to reject the maxims of Moses, Lycurgus, and the patriarchs: we have, by our constitutions of government and the preposterous use made of the doctrines of *equality*, stripped *old* men of their dignity and *wise* men of their influence, and long, long are we to feel the mischievous effects of modern policy.[48]

Webster's assessment ultimately linked the divisive and destructive controversies "in politics as well as medicine" to a range of issues—the repudiation of classical and biblical wisdom, and the folly of the Constitution. Ever the historian, his implicit remedies called for a return to the Spartan simplicity of ancient virtue and the reinstitution of the stern reforms of ancient lawgivers.

The 1790s left Webster a bitter man, callous toward the very system that he had once championed. Disillusioned and a bit defeated, Webster relinquished editorial control of the *Minerva* in 1798 and retreated to New Haven, where he could grumble about the direction of the republic with his similarly disaffected New England friends, such as Timothy Dwight. After a brief stint in state politics, Webster completely retired from public life, and commenced his life as a writer and lexicographer. It was productive work for Webster, but his opinions about the American republic and its citizens were never the same. "Republics are proverbial for *factions*," he later wrote in the entry for "faction" in his dictionary.[49] The definition expressed a lifetime's frustration.

The contagionist faction also troubled Mitchill. In 1799, Mitchill, along Drs. James Tillary and John Rodgers, all members of the Medical Society of the State of New York, formed a committee with the goal of determining the cause of the yellow fever epidemic that devastated New York the preceding fall. The trio scrutinized the available evidence, going so far as to interrogate witnesses. Of course, they concluded the yellow fever arose from domestic sources, but they also claimed that the contagionists fabricated evidence in an attempt to deceive the people. With little more than vague suspicions, the contagionists irresponsibly and maliciously disseminated dangerous, misleading falsehoods. It all began with preconceived notions about the cause of yellow fever. "In a commercial city," the committee members began, "it was natural to suppose, because it has not been thought unreasonable to believe that it *might* be imported from some foreign place: —Accordingly stories were invented and circulated with great diligence, that the *seeds* of this dreadful distemper, imported in the first instance, from foreign shores, had been landed among us." The contagionists specifically implicated the ship *Olive* and the sloop *Iris*, which arrived in New York from Martinique in late June and allegedly introduced yellow fever into the city. Upon further investigation, the members of the committee reported that "these stories were founded in misrepresentation, and fostered by prejudice," and that they sprang from "sources of undiscerning credulity—from motives of self-interest—or, from principles of moral turpitude."[50] To substantiate their accusations, the authors furnished affidavits from the ships' inspectors, who agreed that nobody onboard the *Olive* or the *Iris* had yellow fever, though at least one passenger died from an unidentified disease shortly after arriving.[51]

For Mitchill, the intrigues of the contagionists were another example of the duplicity of early republicans and the danger of factions. In his Fourth of July oration of 1799, Mitchill spoke to such fears. He warned his audience about "strife of parties," but he also connected the proliferation of factions with deeper social anomie. Discussing the decline of American virtue from the Declaration to the Constitution, Mitchill lectured: "Sentiments of distrust and malevolence had gone so far as to disfigure society with . . . the ferocious air of barbarism. Had these ideas been carried further, they must have terminated in the lawless and capricious liberty prevalent among savages, or the sullen, worthless and perfect independence enjoyed by hermits in their caves."[52] The tumult of 1790s, the strife of parties, represented the disintegration of the gentlemanly sociability and charitable spirit that were supposed to form the foundations of public life in the American republic.

Mitchill traced the malfunction to its emergence in the makeup of the republic. Like Webster, he indicted the excess of liberty, which promoted selfishness and "high notions of personal importance," and he specifically condemned the abundance of "perversion and misrepresentations" that circulated through the free press:

> That freedom of speech and of the press so much contended for in republican governments, is employed more than half the time in uttering and disseminating falsehoods of various sorts. Fabrications constantly mislead and perplex the mind. Misstatements beguile and lead astray even those who are serious seekers after truth. The contriving and spreading false news, grows to be a considerable branch of business. . . . What adds to the mass of misrepresentation is, that to defeat the object of one set of lies, there must be an equal quantity of counter-lies put in circulation. The consequence of which is, that when any thing is heard, the first impression it makes is that of falsehood; to be received as true, further proof is required. Thus lying is the rule; and a solitary truth now and then, forms the exception.[53]

Though Mitchill's comment focused on political wrangling, his commentary seemingly identified the degeneration of the yellow fever debate as inevitable. For how could one possibly trust anyone's opinion about anything in a distrustful atmosphere where lying was the rule and truth the exception? Mitchill's rant cast human agents in a tragic drama wherein misinformation beguiled even "serious seekers of truth," and the innate and well-meaning tendency to correct falsehood with falsehood led ineluctably to communication breakdown and made faction a self-fulfilling prophecy.

The critique offered by localists betrayed deeper anxieties about the directions of American society as a whole. Their fears of the unfettered dissemination of information, of factions flouting public welfare, were the fears of a society that was changing from the virtuous republic that investigators hoped it would be, to a bustling commercial nation, where self-interest ruled the day and even information could be manufactured and sold. Mitchill rendered the point clearly in his Independence Day oration. "Contriving and spreading false news," he noted grimly, "grows to be a *considerable branch of business*" (my emphasis).[54] But the question of what was to be done, what balance struck, was still undecided. There were indeed many uncertainties, even contradictions, at the heart of life in the early

American republic, a society caught between the old and the new. What is remarkable is the way these uncertainties, anxieties, and contradictions were communicated, not only in politics but in the tone and content of the debate itself: most looked favorably on some commerce, but they hoped to avoid the decadence it brought and they deplored the cities that sustained it; early republicans looked to the past for guidance, but wanted to escape it; they believed that science could reveal untold mysteries and better the human condition, but they feared that it would lead to irreligion and banish God from existence; and, finally, they believed in the necessity and essential goodness of transparent public discourse, but they lamented what its very openness would permit. The investigators' approaches to solving the riddle of yellow fever had always revealed their attempts to strike balances among the threatening forces in the world—to enshrine these balances as the laws of the land and nature, confirmed by common sense.

There was more than one ingredient involved in the making of the investigator's paranoid style. Rush's singular experiences with persecution reveal the intellectual and religious and intellectual dimensions of the investigators' conspiratorial imaginations. The pattern began early in his life, but it blossomed into full view during the revolutionary crisis of the early 1770s. Like many of his contemporaries, Rush interpreted British economic policies as sure evidence of a concerted plot to rob Americans of their rightful freedoms. Addressing himself to the American public in an essay called "On Patriotism," published in 1773, Rush urged his readers to guard their liberties against the intrusions of the enemy. Calling upon well-worn metaphors of freedom and slavery, he described the invasive British tax measures as "the machinations of the enemies of our country to enslave us by means of the East-India Company." Rush warned against the impending arrival of a number of vessels "freighted to bring over a quantity of tea taxed with a duty to raise a revenue from America." His short essay encouraged Americans in the name of liberty and true patriotism to resist the British plans by refusing to purchase that tea. Rush went so far as to encourage Americans to prevent the landing of the tea, though by what means he never specified. "The baneful chests contain in them a slow poison in a political as well as physical sense," he warned. "They contain something worse than death—the seeds of SLAVERY."[55]

The outbreak of the Revolutionary War by no means curtailed the prevalence of factional intrigue within the United States, as least as far as Rush was concerned. The revolution unleashed the forces of faction, and Rush found himself falling victim to numerous plots of malicious intent. As a member of the Assembly of Pennsylvania, Rush came up against the "malice and cunning" of a wicked cabal, including Timothy Matlack, James Cannon, and Thomas Young, the radical arm of the Assembly, which engineered the ratification of the liberal Pennsylvania State Constitution of 1776. With its unicameral legislature, the constitution was intended to reduce "one of the happiest governments in the world" to a mere "mob government," and its legislators, Rush anticipated in a letter to Anthony Wayne, would soon "become like the 30 Tyrants of Athens."[56] Likewise, as an inspector of the American army hospitals during the war, Rush spoke against the corruption of the director general, Dr. William Shippen, whom Rush believed was pilfering from the hospitals' supplies for personal gain. As Rush expected, or claimed to have expected, Congress ignored his complaints out of ignorance or interest. He resigned from his post, a martyr for a noble cause.[57]

Time and again, in public and private correspondence, Rush denounced the intrigues that plagued him. Regardless of the problem or difficulty he encountered, Rush invariably attributed it to the conspiratorial designs of his opponents. Nefarious persons, motivated by obscure, unnamed private interests, tried to block John Witherspoon's nomination to the presidency of the College of New Jersey, Rush's alma mater, and they did the same to Rush when he appeared set to take the professorship of chemistry at the College of Philadelphia in 1768.[58] Religious bigots led by the "faction" of Joseph Reed, the president of the Pennsylvania Assembly, opposed Rush's attempts to attain a charter for what would become Dickinson College. "British agents" operating in the United States defamed the country in the 1780s, sowing discord in the body politic. Unscrupulous deists worked behind the scenes to undermine the sanctity of the Christian religion and pervert the wholesomeness of good republicanism. In France, various factions, particularly the deists, commandeered the revolution, perpetrated regicide, and sank the country into the misery of despotism.[59]

Rush's conspiratorial imagination focused particularly on his own medical career and the numberless, and normally faceless, operatives who worked tirelessly to destroy it. Of the many persecutions he suffered in his professional life, the story of his travails during the great epidemic in Philadelphia in 1793 stands out. During the epidemic, a group of the city's doctors openly criticized Rush's yellow fever treatment—copious bloodlettings and wholesale purges with two diuretics, mercury and jalap. Led by Dr. Adam Kuhn, the coalition favored a far gentler therapeutic regimen, involving cold baths, moderate doses of wine, and the "bark"—that is, cinchona bark, an antimalarial.[60] Rush, of course, quickly determined that his colleagues, motivated by some perverse desire, had combined against him. He especially balked at the allegations of murder that apparently his opponents leveled against him on account of his drastic curative measures.

As Rush bewailed his victimhood, he simultaneously touted the efficacy of his cure, which he hailed as a virtual panacea. As early as September 5, Rush claimed to Julia that his treatment cured 29 of 30 patients. By September that remarkable success ratio had risen to the more fabulous 99 out of 100. He wrote to Julia, "Could our physicians be persuaded to adopt the new mode of treating the disorder, the contagion might be eradicated from our city in a few weeks."[61] By ignoring his treatment, he believed, dissenting physicians not only showed their ignorance and deceit, they also, in effect, perpetrated murder against their hundreds of patients. To Julia on September 15, "Scores are daily sacrificed to the bark and wine." On September 21, they "continue to murder by rule." And only three days later, "The followers of Kuhn still live to administer poison to our citizens."[62] Similar accusations appeared regularly in Rush's communications during the fall of 1793. By late October, near the end of the fever's reign, Rush reckoned that the disciples of Kuhn had "destroyed at least two-thirds of all who have perished by the disorder."[63]

The story of Rush's alleged persecutions during the yellow fever epidemic of 1793, and his heroic perseverance in the face of such difficulties, is a familiar one. Historians and biographers have retold the essential contours of the episode many times, and a few have even called out Rush's embellishments as the overwrought fantasies that they were. Even the normally uncritical and congratulatory Thomas Flexner—whose title for his 1937 book, *Doctors on Horseback*, conveys some sense of the way he romanticized his subject matter—wondered if Rush really believed that his cure saved as many people as he claimed. And in *Bring Out Your Dead*, the classic account of the outbreak, J. H. Powell depicted Rush as a domineering intellectual, unwilling to accept that anyone but himself could be correct, or that anyone could possibly deny the self-evident efficacy of his treatments. Powell not only rejected Rush's grandiose view of his yellow fever cure, he also questioned the basis of his belief in the persecution—"One searches . . . in vain for those accusations of murder Rush insisted the 'confederacy' laid against him," he wrote.[64] Powell, however, also dismissed Rush's intolerance as a quirk in the public-spirited, if not sometimes overzealous, doctor, who figured as the hero in *Bring Out Your Dead* for his selfless devotion to the public good. When viewed against his lifelong history of paranoia, Rush's near-delusional obsession with the supposed intrigues of his opponents, as well as his overvaluation of his own perspectives, appears not so much as a quirk but as a problem that needs to be explained.

Throughout the epidemic, Rush conjured religious imagery to describe his predicament. "My situation for some time past has been in some respects like that of the children of Israel in the wilderness." Religion informed every aspect of his experience with yellow fever in 1793, from his views of causation (discussed in the previous chapter), to his interpretation of his personal, divine mission.

> Remember, my dear creature, the difference between the law and the gospel. The former only commands us 'to love our neighbors as ourselves,' but the latter bids us to love them *better* than ourselves. 'A new commandment I give unto you, that ye love one another, even as *I have loved you*.' Had I not believed in the full import of that divine and sublime text of Scripture, I could not have exposed myself with so little concern, nay with so much pleasure, for five weeks past to the contagion of the prevailing fever. I did not dare to desert my post, and I believed even *fear* for a moment to be an act of disobedience to the gospel of Jesus Christ.[65]

Rush found a particular resonance in the story of David and Goliath. "My method is too simple for them," Rush wrote of his philosophical adversaries. "They forget that a stone from the sling of David effected what the whole armory of Saul could not do."[66] On the one hand, the allusion to David and Goliath jibed with Rush's conception of epistemology, which he shared with many intellectuals of the early republic. Since God was good and wanted humans to be able to understand and prevent the evils that had afflicted them, he had constructed the world with simple laws so that human common sense, a divine faculty, could easily grasp them. Thus, it stood to reason that the simplicity of ideas, like the simplicity of David's weapon, indicated their effectiveness. The story also tended to a view of himself as an embattled yet righteous underdog. Just as the stone from David's sling felled the physically superior Philistine warrior, so would Rush's divinely inspired cure overcome the wickedness and obstinacy of his numerically superior enemies. With God on his side, Rush could overcome anything. In another letter to Julia, Rush likened his ally, Dr. Griffiths, one of the few doctors who used Rush's treatment, to Joab, a lieutenant of David's, who assassinated rival claimants to the throne. Finally, in tendering his resignation from the College of Physicians in a note to John Redman, Rush once more used the metaphor of David to illustrate his position: "Well might David prefer the scourge of a pestilence to that of the evil dispositions of his fellow men."[67] Here, his reference alluded to a biblical story, later told by James Tytler, in which David, being faced with his choice of punishment for conducting a census of the Israelites, chose three days of plague over three months of persecution and three years of famine.[68]

The romanticization of himself as a David—a humble and godly warrior for truth, fighting the combined treachery of his enemies—suggests that Rush openly embraced, even celebrated, his role. Those whom he most admired also endured prejudiced condemnation. In the midst of his trials, Rush recalled that Thomas Sydenham, the famous and influential English physician, sustained vicious criticism on account of his medical opinions.[69] More importantly, Rush knew that through suffering the torments of persecution, he could identify with Jesus Christ. "The first sermon we find that ever he [Jesus] preached," Rush wrote, "he tells his followers that they should be blessed when all manner of evil was spoke against them."[70] The examples of Sydenham and especially Jesus taught Rush that persecution was to be expected for doing right. Writing to Noah Webster in 1789, Rush enjoined his friend, "Continue to do all the good you can by enlightening your country. *Expect* to be persecuted for doing good, and *learn* to rejoice in persecution."[71] So resolutely had Rush accepted the notion that persecution followed right action that he even seemingly courted it. Only months later, in a letter to John Adams, Rush elaborated on his glorified view of persecution. "I learned . . . from you to despise public opinion when set in competition with the dictates of my judgment or conscience," he began. "So much did I imbibe of this spirit from you that during the whole of my political life I was always disposed to suspect my integrity if from any accident I became popular with our citizens for a few weeks or days."[72]

Eventually Rush achieved some redemption for the persecutions he suffered. After sustaining attacks from William Cobbett, an English émigré and pro-British political writer, Rush sued the "Porcupine" for libel. Here at least Rush had a legitimate gripe. Allying himself with the federalists, Cobbett attacked Rush's republican politics and his sanguinary medical practices. He indicted Rush as a quack, calling him "Dr. Sangrado" after a character in Alain-Rene Lesage's picaresque *Gil Blas* who believed bloodletting cured all bodily ills. Rush won the lawsuit in 1799, forcing Cobbett out of the country. To Rush, the episode provided another example of politics intruding in medicine.[73]

Rush also faced persecution for his opinions about the local origins of yellow fever. Hostility to his opinions grew steadily over the course of the epidemic period. By 1797, the persecutions from his professional colleagues and the public at large had grown so intolerable that he contemplated moving his family to New York, which he believed was more amenable to localism. Philadelphia, his beloved city, had turned against him. "Ever since the year 1793 I have lived in Philadelphia as in a foreign country," Rush confided to Dr. John R. B. Rodgers, in October 1797.[74] Rush repeatedly bemoaned the "calumnies" and "execrations" that his peers heaped upon him for his well-intentioned pursuit of scientific truth.[75] The sting of these reproaches stayed with Rush for the rest of his life. To John Adams, Rush confided that his advocacy of localism was one of the six things he did in his life to incur the malice of his fellow man, but it was clearly the most painful. Recounting the "the folly, ignorance, falsehoods and malice" of his "enemies," Rush claimed, "were I to detail to you the many acts of unkindness, ingratitude, treachery, malice, and cruelty I have received . . . you would wonder how I have survived them." He continued dramatically, "It has been the hatred not of men but of beings actuated by a spirit truly demoniacal."[76] In 1806, having resigned his position from University of Pennsylvania, Rush retired from public life and moved permanently to his house in the country, which he appropriately named "Sydenham."

Rush's ordeals show that he internalized a binary view of the world, born of his particular religious imagination, which filtered its contents in strict terms of right and wrong, and good and evil. Since Rush, by the grace of God, was always both right and good, then his opposition must always be not only wrong but also evil. Hence, his derogation of all medical, as well as political and religious, dissenters to the status of "factions," whose perspectives ought not to be trusted. The same religious outlook also conveniently excused Rush from any critical self-evaluation. Indeed, rather than question his own points of view, Rush found solace for his persecution, or unpopularity, in a depiction of himself as a Christ-like figure, selflessly facing the trials that the righteous had to endure. Though he might suffer now, Rush thought that time and superior wisdom would ultimately vindicate his ideas, just as they had those of his heroes.[77]

Rush's experiences with persecution show that the investigators' conspiratorial imaginations were shaped by ideas about natural inquiry buttressed with the power of religion. In the form of common sense, religion exerted a powerful influence over the disease thought of the investigators, especially the localists, raised on the traditions of Scottish thought. Common sense fired their self-righteousness; it validated their assertions and intuitions. It figured in their historical analyses, in their chemical reveries, and in their efforts to locate a natural theology of disease. Common sense confirmed their causal theories as certainly as it did the existence of God and the accuracy of scriptures. Perhaps then it only followed that they would view their philosophical opponents as they viewed religious heretics, as wayward souls who had strayed from the path of righteousness. Common sense did not license opposing viewpoints, because if truth in science could be had with common sense, which by its very nature ought to be self-evident, then everyone had to agree, or else common sense was not so very common at all, and perhaps therefore nonexistent.

Did the same rationale apply to politics? Could this possibly explain why political discourse likewise degenerated into suspicions of conspiracy? In one of its many incarnations, common sense figured in political thought, and many early republican intellectuals sought to elevate politics to the status of science. Madison certainly believed he had discovered a "science of politics."[78] Rush even more clearly linked politics with science. "There is an indissoluble union between moral, political and physical happiness," Rush claimed in his *Lectures on Animal Life*. With its emphasis on stimuli, his medical system held that republicanism and Christianity constituted the most salutary stimuli for the human body. "If it be true, that elective and representative governments are most favourable to individual, as well as national prosperity, it follows of course, that they are most favourable to animal life."[79] Proper republicanism (*his* republicanism) healed the body. It followed for him that bad politics was like a disease that could infect people, and that factionalism might be a symptom of that malady. Rush's comments show that in the heady days of Enlightenment the lines between science and politics faded. While they do not prove that others thought similarly, they do call for a closer consideration of the way that scientific concepts shaped politics, just as the tone and organization of political discourse helps us understand why scientific debates took the forms they did.

The localists' reasoning showcases the totalizing effects of Enlightenment. By imbuing everything—every question, every problem, and every field of inquiry—with the status of objectivity, it effectively marginalized anyone and anything that fell outside of one's definition of correctness. The Enlightenment created a monster, as many of its critics have pointed out over the years, indeed since the Enlightenment itself.[80] The localists from Rush to Mitchill participated in this counter-Enlightenment backlash by trying to restrict access to enlightened thought, by constraining it to their definitions. Their accusations of conspiracy were anxieties fueled by a perceived excess of Enlightenment—the belief that unwarranted, unsubstantiated, *unenlightened* ideas circulated too freely, and that what they needed were leaders to control and direct discourse.

Despite their diminishing popularity, contagionists never condemned the structures of public discourse, nor unabashedly accused localists of lying, though their opinion hardened and they increasingly called attention to the theoretical roots of localism. Harried by localists, in 1798, Currie denied for the first time that local conditions had any influence on yellow fever— "That the disease was propagated . . . in Philadelphia by specific contagion, which had no connection with the effluvia from putrefying materials, I am warranted in asserting."[81] In the same year, he collected and published a compendium of yellow fever commentary from the most illustrious and oft-cited writers on the disease with the object of furnishing readers with the pure truth, "free from the perversions of salacious and misleading theory, or the misrepresentations of uncharitable, and distorting party spirit."[82] By 1802, when Currie and Cathrall offered their final treatise, their objections had grown more vociferous in proportion to their elevating sense of hopelessness and ineffectuality in the face of localist ascendancy. In their final exasperated plea, Currie and Cathrall urged readers once more to accept the superior factual basis of contagionism, but also to beware of the "sophistry" of localists. But rather than undertaking a point-by-point analysis of localist claims, Currie and Cathrall aimed more broadly to delegitimize localism by reminding their readers of the "reveries" and "abstract speculations"—the assumptions, intuitive leaps, and all-out guesswork—which held it up, and by comparing it unfavorably to the unadorned honesty of contagionism, which all along rested on simple facts.[83]

What made it all so unbearable for Currie, Cathrall, and the rest of the contagionists was not the localists' dishonesty necessarily, but their success, and especially their arrogant conduct—the aggressive manner in which they confederated together and forced their opinions on the people.

> Those gentlemen who affect to deplore, what they so dictatorially pronounce, the errors of those who believe in the *foreign origin* of the malignant yellow fever, and who charge that doctrine with being 'fatal to the lives and injurious to the property of our citizens,' do not seem to be aware that the charge may be retorted with double force upon themselves. Nor, do they appear to be sensible of the resemblance of their conduct to a certain usurper, who while he was laying waste all before him, and increasing the misery of surrounding nations, professed himself the friend of humanity, and declared he was only contending for the freedom and happiness of mankind.[84]

The thinly veiled allusion to Napoleon Bonaparte proved fitting for the times. Napoleon had only just declared himself consul for life and then invaded Switzerland, and so the comparison enabled Currie and Cathrall to depict their enemies as ruthless, lawless conquerors. The analogy, though, would have had much deeper resonance for people of the United States in late 1802, as its authors well knew. More than another conqueror, Napoleon, the usurper of France and the terror of Europe, had come to symbolize the end of the age of Enlightenment. Napoleon finally and irrevocably ended the French Revolution, and with it the hopes of establishing a peaceable society ruled by reason. After a decade of yellow fever, with defeat all but being acknowledged, Currie and Cathrall found their own situation mirrored in contemporary affairs. They had come to the study of yellow fever believing that reasoned inquiry would lead to truth; but like all those who mourned the loss of France, they found that the ignorance, party interest, and bald power of a faction upset their high hopes.

In retrospect, Currie and Cathrall seem to have had the sounder claim. Although they loudly reproached the machinations of the contagionist faction, localists from Webster to Rush more closely resembled a faction. At the very least, the localists more effectively took advantage of the discursive tools available. They knit themselves into a wide community that spanned the United States and was held together through epistolary exchange, personal acquaintanceships, and unlikely friendships, such as the one that developed between Rush, a democratic-republican universalist, and Webster, the conservative, nationalistic federalist and Congregationalist. Localist ideas enjoyed privileged places in universities such as Columbia and the University of Pennsylvania, where localist professors dispersed their creed. Localists also published pamphlets more prolifically than contagionists, and otherwise fought more tenaciously and more creatively. More to the point, localists brooked no opposition in their quest to define the truth about yellow fever, and *they*, not the contagionists, refused to compromise and establish reciprocal exchanges. For Currie and Cathrall, and the contagionists overseas, it was hard not to conclude that the localists had formed a faction that ruthlessly and self-interestedly usurped fever discourse, and that they perpetuated their ideas through private networks and superior material organization.

The arrogance of the enlightened lay at the root of controversies "in politics as well as medicine." Localists in particular could not accept that they might be wrong or that their opponents might have legitimate points of view, based on good evidence and reasoning. But at least some among them came close. Elihu Hubbard Smith certainly did in autumn 1795, at very nearly the same time he was composing the verses that appear at the beginning of this chapter. Surveying the medical controversy erupting in New York, Smith mused over his own culpability in it. He wrote in his diary, "Do I not see ignorance, pride, stupidity, carelessness, & a superstitious veneration for foreign writers, & a mean jealousy of an illustrious writer of our own country [i.e., Rush], go hand in hand, & as it were, conspire, against the lives of men?" The comment rescued Rush's reputation, and it mocked the contagionists' servile allegiance to British and French medical authorities. Then, in a moment of introspection, Smith turned his critical gaze inward. "Is it a foolish vanity, which deceives me, when I suspect that I am less under the influence of prejudice than many; & that I should venture my reputation freely to do good; & that no bigottry to a peculiar system would prevent my readily yielding any notion, which I had once entertained, if I found it false?" Smith seemed to be edging closer to a recognition of the relativity of yellow fever opinions; to an admission that he was no more free from prejudice than those he accused; and, indeed, to an understanding that the whole matter of prejudice, faction, and conspiracy was nonsense in the first place—just the paranoid expressions of uncertainty and misunderstanding. Instead, he replied tersely to his own question, "I think not."[85] His answer helps us understand why the yellow fever debate took the form it did.

Conclusion

"A New Era in the Science of Medicine"?

In the midst of a yellow fever outbreak in Philadelphia in September 1805, Benjamin Rush wrote to his friend John Adams. The act itself was not an unusual one for the aging doctor (he would turn sixty in a few months)—he was a devoted letter-writer, as the thousands of letters he composed in his life attest, and he had always taken time even in the midst of epidemics to correspond with family and friends. But on this particular occasion, something new had come over Rush. Prospects had brightened considerably, and Rush's letter showed it. Gone were the sleepless nights, the endless house visits to patients, the fervent devotions to God, and even, to Rush's delight, the worst of the "persecutions." All the demons that had haunted him throughout the epidemic period seemed to have vanished. And, so, he wrote exuberantly to Adams of the changes that had come over the United States and its medico-scientific community:

> A new era has begun in the science of medicine in our city since the appearance of the yellow fever among us. No channel has as yet been discovered through which it could have been conveyed to us from a foreign country;

and what is more against its importation, no one of the persons who have been infected by the foul air emitted by a large bed of putrid oysters in Southwark and who has sickened or died in the city, has propagated the disease. Many of our citizens of the second class have been led by these facts to believe . . . that it is not contagious.[1]

Rush's enthusiasm betrayed his optimism about the success of localism and the fate of yellow fever in the early American republic. Rush and the localists certainly had not finally and definitively determined the cause of yellow fever (nascent public health bodies in the American cities still warily maintained quarantine procedures, for example), but they had settled the debate among natural philosophers. Writing from across the Atlantic in 1801, the contagionist John Haygarth lamented that the tide of American public opinion had already turned in favor of localism: "In the newspapers, in conversation, and in letters from America, it is asserted with great positiveness so as to obtain general belief."[2] By 1805, hardly a contagionist could be found who openly advocated for the doctrine, and the "citizens of the second class," as Rush condescendingly referred to them, finally embraced the localist explanation. Rush even flattered himself by thinking that William Currie, the bulwark of the opposition, changed his mind.[3] Currie had not, but he had more or less retired from the arena of yellow fever commentary with his final, resentful essay of 1802.

Always the more prominent and well-connected intellectuals, localists chased contagionists from the battle and commandeered public fever discourse. Their pamphlets circulated in city bookstores and catalogues, their creed thrived in the localist-controlled *Medical Repository*, and for a time at least they expounded the word of localism to their students, many of whom, like Phineas Jenks and Charles Caldwell, became vocal supporters. Their voices drowned out those of the competition, but it was not brute force alone that won the debate. Localist arguments better appealed to common sense. They reconciled the facts of the yellow fever epidemics with the existence of locally situated miasmas; they showed that its occurrences conformed to patterns of disease occurrences that had swept across the globe in ages past; they explained how the fermentation of noxious materials emitted miasmas of yellow fever, and some even claimed to have identified the precise chemical structure of those yellow fever particles; and, finally, localists rationalized the natural generation of miasmas from filth as an expression of God's purpose, written into creation. Looking from the position of the

1790s and early 1800s, localism would have appeared as vibrant and compelling as any other well-accepted scientific notion.

Contagionists perhaps did not deserve the fate they received. Their insistence that Americans imported the contagious disease from abroad reflected a careful consideration of the facts (a consideration that localists could only truly dismiss by accusing contagionists of intentionally lying). Their problem was not one of evidence or methodology, but from their own philosophical restraint and modesty. With the exception of the eccentric James Tytler, contagionists refused to insist on knowledge about things they could not see or feel, and they avoided the fanciful and indulgent speculations for which they criticized localists. They never claimed to know the properties of their contagious particles or the processes they produced. That left others wondering when in the folds of time contagious particles came into being (did they agree with Tytler?), and why they would exist at all. Contagionists also proved more accommodating than their counterparts. They confessed that local environmental conditions nourished contagions, allowing them to develop into epidemics. That admission could have established a more peaceful and productive compromise, but the localists would not have it. Only time vindicated the contagionists' ideas.

Happily for everyone, as Rush also suggested in his letter to Adams, yellow fever had abated considerably since the devastation of 1798. As the Haitian Revolution intensified, Americans gradually lost contact with the war-torn colony of Saint-Domingue, their major trading partner for sugar and coffee, and the source of their annual epidemics. By 1805, when the United States under Thomas Jefferson officially severed commercial ties with independent Haiti, the revolution had already come to an end and, without a regular influx of nonimmune Europeans, so too had the yellow fever epidemics.[4] Minor outbreaks in Philadelphia and New York in 1805 were the last of the epidemic period. The disease reemerged with a vengeance in Baltimore in 1819, killing more than 2,000, but just as quickly retreated. A final outbreak in Philadelphia in 1820 took 83 lives, and New York's last epidemic in 1822 claimed about 250 more. For the rest of the nineteenth century, yellow fever became a problem of the great southern cities, particularly New Orleans, but it would never again strike the seaport cities of the northeastern United States.[5]

The end of yellow fever also marked the end of the intellectual ferment that accompanied it. A few treatises trickled from the presses after 1805 but then abruptly stopped. As for the investigators, most of them simply abandoned the pursuit of yellow fever as suddenly as they had gotten into it. Neither Rush, Mitchill, or Webster ever wrote about the disease again; nor did Isaac Cathrall, Charles Caldwell, or James Tytler. Of all the investigators, only one, Felix Pascalis Ouvière, ever produced another work about the disease again, and not until its brief reemergence in 1819.[6] That was fitting though. For the bulk of investigators, yellow fever inquiry never constituted a professional occupation or obligation. A gentlemanly pursuit, fever inquiry constituted something like a civic duty for the learned and concerned, who believed that public discourse over matters of common concern could shape and mold perceptions of scientific knowledge as well as civic and godly duty.

But to return to Rush's letter of 1805, there was a matter about which he was quite mistaken—"a new era in the science of medicine" had most certainly not begun, though one was seemingly coming to an end. The age of common-sense science did not finally and irrevocably end with the fall of yellow fever, but its luster faded in the coming years. Like Rush, the latest generation of Philadelphia, Princeton, and New York intellectuals educated in Scottish philosophy in Edinburgh and Aberdeen were aging and dying. And with the Atlantic unfit for crossing due to the upheavals of the Napoleonic Wars, connections between the United States and Scotland gradually dissolved.[7]

The various strands of science that the investigators pulled together during the epidemic period went in their own directions too. The broad observational empiricism that undergirded the historical study of yellow fever manifested itself more fully in the pursuits of nineteenth-century natural historians, who scoured the American countryside for facts about its plants, animals, and minerals, hoping that they might one day lead to some hidden truth about their environments.[8] In chemistry, the laboratories of Benjamin Silliman and Robert Hare tempered the enthusiasm that greeted the appearance of Lavoisierian concepts in the 1790s, although the investigators' speculative zeal, the chemical sublime, survived in the romantic era of American science. Of all the fields of inquiry brought to bear on the problem of yellow fever, only natural theology survived virtually unchanged to the present in the form of creationism or intelligent design (hardly a reassuring sign of its explanatory power!).

Change occurred even more dramatically in the areas of medical education and practice, broadly conceived. At the conclusion of the Napoleonic

Wars, the University of Edinburgh, with its emphases on rationalistic physiological systems, no longer attracted the same numbers of Americans. Those traveling overseas for their educations increasingly chose the rigorous clinical settings of the Paris hospitals, which offered training in bedside observation and autopsy. There they would look into the bodies of their patients and trace disease into the very tissues in which they were embedded. French medical ideas soon came to dominate in the upstart medical schools and journals of the antebellum United States.[9] The debate between localists and contagionists again flared up when yellow fever settled in the American South, and especially when cholera, the dread disease of the nineteenth century, began to afflict the American cities, but the subsequent generations of investigators were more inclined to look for answers in the bodies and corpses of their patients as much as in ships and putrid effluvia.[10]

With these broad changes in the ways that inquirers disciplined their fields of inquiry came changes in the spaces where they offered and contested natural knowledge. Science increasingly moved out of the public sphere. Behind the closed doors of hospitals, laboratories, exclusive and professional associations, and indeed public health bodies, professionals could impose a kind of discipline on their subjects that the public arena did not accept. Some of these changes were under way in the late 1790s, with the rise of municipal health boards and health laws. Yellow fever itself influenced broad changes in the trajectory of American science and medicine. The yellow fever ferment exposed the fragility of common-sense inquiry. By elevating one's impressions and convictions to the status of truths, common sense fostered intolerance toward opposing viewpoints, which became something more like heresies than contrary points of view. The debate also revealed the limitations of public science; publicity itself dissolved the debate into a controversy and turned investigators against the open discursive arena in which they contested yellow fever.

But for all its impermanence, the epidemic period still marked an incredible period of productivity. By bringing it back to light, by reconstructing the content of the yellow fever ferment, I have tried to resurrect perspectives on nature and epistemology, as well as beliefs about diseases and their causes that have all long since fallen out of favor among doctors and scientists, and which may now only appear strange, even incomprehensible, to the modern eye. In doing so, I hope that I have dispelled some of that strangeness and incomprehensibility, and perhaps corrected some of the condescension with which we too frequently view old science. The contest between contagionists and localists did not pit stale, unchanging theories against each other, an impression given by much of the scholarship on disease thought; rather, it challenged inquirers to wield growing and shifting domains of scientific knowledge against old questions.

The success of localism reveals the high appeal of common-sense science during the early years of the American republic. Both localism and the common-sense approach to scientific knowledge-making rested on core axioms: God created the world with purpose, and as part of that purpose he gave human beings a common sense that innately grasped reality and truth; therefore, truths would appear as plausible and consistent elements of a coherent, designed world. Common sense thus explains salient features of the era's natural inquiry. It accounts for Americans' fondness for natural history, a field of knowledge that invited observers to go out and look for God in nature. It explains how intellectuals of the time unproblematically united seemingly disparate fields of inquiry; how they used theology to illuminate nature and vice versa, and how science bled into politics (for did not a common thread run through them all?). Common-sense views allowed inquirers to reconcile science and religion, and to escape the clutches of the godless minions who wanted to bend nature to their own wills and to pervert society. It probably also explains why natural philosophers failed to achieve much of any lasting scientific value. Through the lens of common sense, inquirers could see what they wanted to see.

We might wonder how far such ideas about purpose and disease inquiry lasted in the next century. Did nineteenth-century disease investigators who preferred localism also find that theory to better jibe with God's ostensible purpose? To speculate even further, did the decline of localism occur simply as a result of the self-evident correctness of the microbiological "germ" theory? Or was it more than a coincidence that the sudden downfall of localism and the rise of microbiology came right after the path-breaking work of Charles Darwin, whose theory of evolution by natural selection radically altered human perceptions of the purposes of things in existence, including the invisible life-forms that would, in short time, achieve near-universal recognition as the agents that caused disease?

The case of yellow fever in the early republic reminds us that disease inquiry and knowledge form parts of our worldviews. What counts as knowledge at any given time is always a part of a constellation of beliefs and assumptions, determined in part by historical contexts. As we face down a host of natural problems—global warming, environmental degradation, reckless population growth—we would do well to make sure that we do not see only what we want to see. And though modern medicine has diminished the threat of disease, the appearance of AIDS and Ebola, and the persistence of malaria, tuberculosis, measles, and even polio in many parts of the world remind us that the scourge of pestilence has not ceased and might one day reemerge. Even where infectious diseases have been subdued, modern diseases, such as cancers, Alzheimer's, and autism, entwined in our genomes, put us in a position at least dimly similar to that of the early republicans who confronted yellow fever. The tools of investigation have changed considerably. But we still hope with Rush that a "new era in the science of medicine" is dawning. Maybe someday it will.

Notes

INTRODUCTION

1. John C. Bugher, "The Pathology of Yellow Fever," in *Yellow Fever*, ed. George Strode (New York: McGraw-Hill, 1951), 141–142.

2. Mathew Carey, *A Short Account of the Malignant Fever, Lately Prevalent in Philadelphia* (Philadelphia: Mathew Carey, 1793), 112–116, compiled a list of church burials, 4,042 in all. His total does not account for the dozens of people buried outside of church grounds, like the hundreds deposited in the city's potter's field, a burial place for the unclaimed. Eve Kornfeld, "Crisis in the Capital: The Cultural Significance of Philadelphia's Great Yellow Fever Epidemic," *Pennsylvania History* 3 (1984), 189, puts the number at 5,000.

3. The classic account is John Harvey Powell, *Bring Out Your Dead: The Great Plague of Yellow Fever in Philadelphia in 1793* (Philadelphia: University of Pennsylvania Press, 1949). J. Worth Estes and Billy G. Smith, eds., *A Melancholy Scene of Devastation: The Public Response to the 1793 Philadelphia Yellow Fever Epidemic* (Canton, MA: Science History Publications, 1997).

4. David K. Patterson, "Yellow Fever Epidemics and Mortality in the United States, 1693–1905," *Social Science and Medicine* 34, no. 8 (1992), 857.

5. Carey, *Short Account*, 34.

6. Stubbins Ffirth, *An Inaugural Dissertation on Malignant Fever; with an Attempt to Prove Its Non-Contagious Nature, from Reason, Observation, and Experiment* (Philadelphia: B. Graves, 1804), 11.

7. The question of what to call these investigators and what they did deserves clarification. "Scientist" is inappropriate because the term itself was not coined until the 1830s, and not regularly used until the twentieth century, when it described professional practitioners operating within well-defined, methodologically rigorous and specific disciplines. "Scientist" imputes a type of social identity to the investigators that they did not possess. The term "natural philosopher" better captures the way investigators thought of themselves, as philosophers who investigated natural problems (among other types of problems), and in doing so often fused together disparate fields of knowledge. On the other hand, the investigators clearly

imagined what they were doing as science—in fact, they referred to many different sciences—united by a broadly empirical and inductive methodology. Thus I use the noun "science" and adjective "scientific." Steven Shapin discusses problems of terminology, albeit in a different context, in *The Scientific Revolution* (Chicago: University of Chicago Press, 1998), 1–6.

8. Oyewale Tomori, "Yellow Fever: The Recurring Plague," *Critical Reviews in Clinical Laboratory Sciences* 41, no. 4 (2004), 391–427.

9. For the story of the discovery of the vector and virus, Michael B. A. Oldstone, *Viruses, Plagues, and History: Past, Present, and Future* (New York: Oxford University Press, 2010), 119–126, 132–133; for yellow fever in the Americas: Jean Slosek, "*Aedes aegypti* Mosquitoes in the Americas: A Review of Their Interactions with the Human Population," *Social Science and Medicine* 23, no. 3 (1986), 249–257.

10. Thomas Apel, "The Rise and Fall of Yellow Fever in Philadelphia, 1793–1805," in *Nature's Entrepôt: Philadelphia's Urban Sphere and Its Environmental Thresholds*, ed. Brian Black and Michael Chiarappa (Pittsburgh: University of Pittsburgh Press, 2012), 55–74.

11. Thomas Kuhn, *The Structure of Scientific Revolutions* (Chicago: University of Chicago Press, 1962), 4. Jan Golinski offers an excellent overview of "constructivism," as he refers to the movement, in *Making Natural Knowledge: Constructivism and the History of Science* (New York: Cambridge University Press, 1998). The exceptional example of this type of work, and a major influence on me, is Simon Schaffer and Steven Shapin, *Leviathan and the Air-Pump: Hobbes, Boyle, and the Experimental Life* (Princeton, NJ: Princeton University Press, 1985).

12. Philip Gould, *Barbaric Traffic: Commerce and Antislavery in the Eighteenth-Century Atlantic World* (Cambridge: Cambridge University Press, 2006), chapter 5; Bryan Waterman, *The Republic of Intellect: The Friendly Club of New York and the Making of American Literature* (Baltimore: Johns Hopkins University Press, 2007), chapter 5; Katherine Arner, "Making Yellow Fever American: The Early American Republic, the British Empire and the Geopolitics of Disease in the Atlantic World," *Atlantic Studies* 7 (December 2010), 447–471; Simon Finger, *The Contagious City: The Politics of Public Health in Early Philadelphia* (Ithaca, NY: Cornell University Press, 2012), 120–162.

13. By contrast, well-known studies of early American science focus on single fields. Natural history has several books. Pamela Regis, *Describing Early America: Bartram, Jefferson, Crevecoeur, and the Influence of Natural History* (Philadelphia: University of Pennsylvania Press, 1999); Susan Scott Parrish, *American Curiosity: Cultures of Natural History in the Colonial British Atlantic World* (Chapel Hill: University of North Carolina Press, 2006); Andrew Lewis, *A Democracy of Facts: Natural History in the Early Republic* (Philadelphia: University of Pennsylvania Press, 2011). On electricity, James Delbourgo, *A Most Amazing Scene of Wonders: Electricity and Enlightenment in Early America* (Cambridge, MA: Harvard University Press, 2006). Delbourgo's fine study ably extrapolates the significance of electricity to scientific study in early America as a whole.

14. Studies of controversies have shown that they uniquely bring ideas and anxieties to surface. Kuhn argued that scientific controversy constitutes the generative force that pushes sciences onward and leads to paradigm changes. See Schaffer and Shapin, *Leviathan and the Air-Pump*; Martin Rudwick, *The Great Devonian Controversy: The Shaping of Scientific Knowledge among Gentlemanly Specialists* (Chicago: University of Chicago Press, 1985). Graham Burnett uses a court case to the same end in *Trying Leviathan: The Nineteenth-Century Court Case That Put the Whale on Trial and Challenged the Order of Nature* (Princeton, NJ: Princeton University Press, 2010).

15. Erwin Ackernecht, "Anticontagionism between 1821 and 1867," *Bulletin of the History of Medicine* 22 (1948), 117–153. His thesis has provoked both support and dissent, although virtually all agree that notions of disease causation and public health responses reflected political concerns. A notable supporter is Richard J. Evans, *Death in Hamburg: Society and Politics in the Cholera Years 1830–1910* (Oxford: Clarendon Press, 1987); dissenters include Andrew Aisenberg, *Contagion: Disease, Government, and the "Social Question" in Nineteenth-Century France* (Stanford, CA: Stanford University Press, 1999); Peter Baldwin, *Contagion and the State in Modern Europe, 1830–1930* (New York: Cambridge University Press, 1999). For the "Ackerknecht thesis" in the United States, see Martin Pernick, "Politics, Parties, and Pestilence: Epidemic Yellow Fever in Philadelphia and the Rise of the First Party System," *William and Mary Quarterly* 29 (October 1972), 559–586. Pernick's essay was reissued in Estes and Smith, *Melancholy Scene of Devastation*, 119–136. In an "Afterword" (136–138), he offers some thoughts on the article's limitations.

16. For overviews of these health reforms, see John Duffy, *The Sanitarians: A History of American Public Health* (Urbana-Champaign: University of Illinois Press, 1990), 38–48; Martin Melosi, *The Sanitary City: Urban Infrastructure in America from Colonial Times to the Present* (Baltimore: Johns Hopkins University Press, 2000). Finger's *Contagious City*, 120–162, presents a more a delicately nuanced view of the politics of public health in early Philadelphia.

17. Noah Webster, *Brief History of Epidemic and Pestilential Diseases* (Hartford, CT: Hudson and Goodwin, 1799), viii.

18. Lewis, *Democracy of Facts*, esp. 13–45.

19. For common sense as a model of scientific inquiry, see David B. Wilson, *Seeking Nature's Logic: Science in the Scottish Enlightenment* (University Park: Pennsylvania State University Press, 2009); Theodore Dwight Bozeman, *Protestants in an Age of Science: The Baconian Ideal and Antebellum American Religious Thought* (Chapel Hill: University of North Carolina Press, 1977) argues that common sense bridged the gap between religion and science for faculty and students at Princeton in the nineteenth century. More recently, James Delbourgo considers the ways in which common sense mediated the American reception of Elisha Perkins's metallic tractors in "Common Sense, Useful Knowledge, and Matters of Fact in the Late Enlightenment," *William and Mary Quarterly* 61, no. 4 (October 2004), 643–684.

20. Bozeman, *Protestants in an Age of Science* and Mark Noll, *Princeton and the Republic, 1768–1822: The Search for a Christian Enlightenment in the Era of Samuel Stanhope Smith* (Princeton, NJ: Princeton University Press, 1989); Nina Reid-Maroney, *Philadelphia's Enlightenment, 1740–1800: Kingdom of Christ, Empire of Reason* (Westport, CT: Greenwood, 2001); Lewis, *Democracy of Facts*, 107–128; Leigh Eric Schmidt, *Hearing Things: Religion, Illusion, and the American Enlightenment* (Cambridge, MA: Harvard University Press, 2000); Delbourgo, *Most Amazing Scene of Wonders*.

21. Despite the scarcity of his actual scientific study, Jefferson does loom large in studies. Take, for example, the only real survey of early republican science, John C. Greene, *American Science in the Age of Jefferson* (Ames: Iowa State University Press, 1984); also works on natural history, such as Regis, *Describing Early America*; and Lee Alan Dugatkin, *Mr. Jefferson and the Giant Moose: Natural History in Early America* (Chicago: University of Chicago Press, 2009). Rather than thinking of "American science in the Age of Jefferson," it would be more accurate to discuss "science in the age of Rush," for it was his approach to and sensibilities about science that prevailed in the minds of practitioners of scientific inquiry, and in young nation's universities and public institutions.

22. An excellent survey of these developments, with essays devoted to major disciplines, is David Cahan, ed., *From Science to the Sciences: Writing the History of Nineteenth-Century Science* (Chicago: University of Chicago Press, 2003).

CHAPTER I

1. Benjamin Rush to Enoch Green, 1761, in *The Letters of Benjamin Rush*, ed. Lyman H. Butterfield (Princeton, NJ: Princeton University Press, 1951), 1:3.

2. Mathew Carey, *A Short Account of the Malignant Fever, Lately Prevalent in Philadelphia*, 2nd ed. (Philadelphia: Mathew Carey, 1793), 18.

3. *Gazette of the United States*, December 11, 1793. Penn's ghost was not incorrect about Penn's intentions. See Simon Finger, *The Contagious City: The Politics of Public Health in Early Philadelphia* (Ithaca, NY: Cornell University Press, 2012), 7–20.

4. Rush to John Morgan, November 16, 1766, in *Letters of Benjamin Rush*, 1:26. Morgan had only just founded the College of Philadelphia's medical school.

5. Biographical information about Rush appears in many sources. For a thorough treatment of his many activities, see Nathan G. Goodman, *Benjamin Rush: Physician and Citizen, 1746–1813* (Philadelphia: University of Pennsylvania Press, 1934), 1–42; for Rush as domineering intellectual, see Michael Meranze in *Laboratories of Virtue: Punishment, Revolution, and Authority in Philadelphia, 1760–1835* (Chapel Hill: University of North Carolina Press, 1996).

6. Information about Currie is hard to come by. For biographical details, see College of Physicians, *Transactions of the College of Physicians* (Philadelphia: Thomas Dobson, 1887), clix–clxi.

7. George Carter, *An Essay on Fevers, Particularly on the Fever Lately so Rife in South Carolina* (Charleston, SC: W. P. Harrison, 1796). David Ramsay, "Extract from an Address delivered before the Medical Society of South-Carolina," *Medical Repository* 4, no. 1 (January 1801), 98–103; Joseph Mackrill, *The History of the Yellow Fever* (Baltimore: John Hayes, 1796); Gentleman of the Faculty, *Observations on Doctor Mackrill's History of the Yellow Fever* (Baltimore: John Hayes, 1796); John Beale Davidge, *A Treatise on the Autumnal Endemial Epidemick of Tropical Climates, Vulgarly Called the Yellow Fever, Containing its Origin, History, Nature, and Cure* (Baltimore: W. Pechin, 1798).

8. For the roots of contagionist thinking, Charles-Edward Amory Winslow, *The Conquest of Epidemic Disease: A Chapter in the History of Ideas* (Princeton, NJ: Princeton University Press, 1943), 75–87; an insightful work by Susan Sontag argues that all diseases are treated metaphorically as contagious, even if they are not. See her *Illness as Metaphor and AIDS and Its Metaphors* (New York: Picador, 2001). For the plague of Athens, Thucydides, *The History of the Peloponnesian War*, 2.48; Leviticus 12–15.

9. Winslow, *Conquest of Epidemic Disease*, 131–136.

10. Richard Bayley, *An Account of the Epidemic Fever which Prevailed in the City of New York*, 38.

11. Isaac Cathrall, *A Medical Sketch of the Synochus Maligna, or Malignant Contagious Fever* (Philadelphia: Thomas Dobson, 1794), 10–13.

12. As an example, Rush admitted the contagiousness of smallpox throughout his life. See Benjamin Rush, *Observations upon the Origin of the Malignant Bilious, or Yellow Fever in Philadelphia* (Philadelphia: Budd and Bartram, 1799), 11–12. The standard source for smallpox in the United States is Elizabeth Fenn, *Pox Americana: The Great Smallpox Epidemic of 1775–1783* (New York: Hill and Wang, 2001).

13. For the connection between localism and endemic malaria in Greece, see Roy Porter, *The Greatest Benefit to Mankind: A Medical History of Humanity from Antiquity to the Present* (New York: W.W. Norton, 1997), 60; William McNeill, *Plagues and Peoples* (New York: Anchor Books, 1977), 115–118.

14. On malaria in colonial North America, see Margaret Humphreys, *Malaria: Poverty, Race, and Public Health in the United States* (Baltimore: Johns Hopkins University Press, 2001), 20–29; and J. R. McNeill, *Mosquito Empires: Ecology and War in the Greater Caribbean, 1620–1914* (Cambridge: Cambridge University Press, 2010), 198–202, 203–207.

15. David Ramsay, *Dissertation on the Means of Preserving Health, in Charleston, and the Adjacent Low Country* (Charleston, SC: Markland & M'Iver, 1790), 19; William Currie, *A Dissertation on the Autumnal Remitting Fever* (Philadelphia: Peter Stewart, 1789). Currie dedicated the treatise to Benjamin Rush, who was to become his rival during the yellow fever years, and noted that it was a "testimony of the very exalted Opinion which the Author entertains, of his [Rush's] amiable and engaging Manners as a Gentleman, and of his distinguished Abilities in the Several Departments of Science, and especially in that of Medicine." See the dedication.

16. Much more will be said of the chemical construction of yellow fever in Chapter 3. For an introduction to the topic, see Alain Corbin, *The Foul and the Fragrant: Odor and the French Social Imagination*, trans. Miriam Kochan (Cambridge, MA: Harvard University Press, 1986), 1–56.

17. F.O.P., "Appendix," in Thomas Condie and Richard Folwell, eds. *History of the Pestilence, Commonly Called Yellow Fever, which Almost Desolated Philadelphia, in the Months of August, September & October, 1798* (Philadelphia: Richard Folwell, 1799), xiii–xv.

18. Samuel Johnson, *A Dictionary of the English Language* (Dublin: 1775), unpaginated, find under the heading "fact."

19. The phrase is borrowed from Jessica Riskin, *Science in the Age of Sensibility: The Sentimental Empiricists of the French Enlightenment* (Chicago: University of Chicago Press, 2002), 142. On the history of the fact, Barbara Shapiro, *A Culture of Fact: England 1550–1720* (Ithaca, NY: Cornell University Press, 1999); Mary Poovey, *A History of the Modern Fact: Problems of Knowledge in the Sciences of Wealth and Society* (Chicago: University of Chicago Press, 1998).

20. For the fact culture of the early republic, see especially Andrew Lewis, *A Democracy of Facts: Natural History in the Early Republic* (Philadelphia: University of Pennsylvania Press, 2011). For the place of the fact in the knowledge culture of the early republic, James Delbourgo, "Common Sense, Useful Knowledge, and Matters of Fact in the Late Enlightenment," *William and Mary Quarterly* 61, no. 4 (October 2004), 643–684.

21. Lewis, *Democracy of Facts*; Steven Shapin discusses the social bonds that sustained natural inquiry in *A Social History of Truth* (Chicago: University of Chicago Press, 1994).

22. Cathrall, *A Medical Sketch of the Synochus Maligna*, 3.

23. Richard Bayley, *An Account of the Epidemic Fever which Prevailed in the City of New York, during Part of the Summer and Fall of 1795* (New York: T. and J. Swords, 1796), 6, 13.

24. David de Isaac Cohen Nassy, *Observations on the Cause, Nature, and Treatment of the Epidemic Disorder, Prevalent in Philadelphia* (Philadelphia: Parker & Co., 1793), 11, 13.

25. William Currie, *A Treatise on the Synochus Icteroides, or Yellow Fever; As It Lately Appeared in Philadelphia* (Philadelphia: Thomas Dobson, 1794), vi–vii.

26. Virtually all recent scholars of the Scientific Revolution deny that it was a single, unified phenomenon with a clear, unambiguous research program. See Steven Shapin, *The Scientific Revolution* (Chicago: University of Chicago Press, 1996); Lisa Jardine, *Ingenious Pursuits: Building the Scientific Revolution* (New York: Anchor Books, 2000). Thus, the elevation of empiricism and inductivism, while certainly one of the legacies of the Scientific Revolution, was not its central contribution. Indeed, some of the key works of the period, like Newton's *Principia Mathematica*, which the investigators greatly admired, could not be described in that way.

27. Hippocrates probably existed, but the works that bear his name were undoubtedly written by many people, the students of Hippocrates. Andrew Cunningham, "The Transformation of Hippocrates in Seventeenth-Century Britain," in *Reinventing Hippocrates*, ed. David Cantor (Burlington, VT: Ashgate, 2002), 102–104.

28. On the spread of neo-Hippocratic thought, James C. Riley, *The Eighteenth-Century Campaign to Avoid Disease* (Basingstoke, UK: Macmillan, 1987). For Cullen specifically, see W. F. Bynum, "Cullen and the Study of Fevers in Britain, 1760–1820," *Medical History*, Supplement No. 1 (1981), 135–147.

29. William Hillary, *Observations on the Changes of the Air and the Concomitant Epidemical Diseases* (London: 1759); Robert Jackson, *A Treatise on the Fevers of Jamaica* (London: J. Murray, 1791); and John Hunter, *Observations on the Diseases of the Army in Jamaica* (London: J. Nicol, 1788).

30. The lineaments of the book trade and its importance specifically on medical subjects are best treated by Richard Sher, *Enlightenment and the Book: Scottish Authors and Their Publishers in Eighteenth-Century Britain, Ireland, and America* (Chicago: University of Chicago Press, 2009), esp. 503–596. For the trans-Atlantic dimensions of yellow fever discourse, see Katherine Arner, "Making Yellow Fever American: The Early American Republic, the British Empire and the Geopolitics of Disease in the Atlantic World," *Journal of Atlantic Studies* 7, no. 4 (December 2010), 447–471.

31. Shapin has much to say about epistolarity and the development of social bonds among natural inquirers in *Social History of Truth*.

32. For the influence of the University of Edinburgh, see Deborah Brunton, "The Transfer of Medical Education: Teaching at the Edinburgh and Philadelphia Medical Schools," in *Scotland and America in the Age of the Enlightenment*, 242–258; J. Rendall, "The Influence of the Edinburgh Medical School on America in the Eighteenth Century," in *The Early Years of the Edinburgh Medical School*, ed. R.G.W. Anderson and A.D.C. Simpson (Edinburgh, 1976), 95–124. For works on American students who studied at Edinburgh, see Helen Brock, "Scotland and American Medicine," in Helen Brock and William Brock, *Scotus Americanus: A Survey of the Sources for the Links between Scotland and American in the Eighteenth Century* (Edinburgh: University of Edinburgh Press, 1982), 114–126. In her Appendix, Helen Brock lists all individuals she could find who studied at Edinburgh and then practiced in the thirteen colonies. There were hundreds.

33. As with the other neo-Hippocratics, Rush's relationship with Hippocrates was conflicted—he knew Hippocrates's ideas were quite flawed but nevertheless celebrated his methods, especially as they were distilled by Sydenham. Carl J. Richard, *The Founders and the Classics: Greece, Rome, and the American Enlightenment* (Cambridge, MA: Harvard University Press, 1994), 202–204, 212.

34. Samuel Bard, *A Discourse upon the Duties of a Physician* (New York: A and J Robertson, 1769), 6.

35. College of Physicians, *The Charter, Constitution, and Bye Laws of the College of Physicians of Philadelphia* (Philadelphia: Zachariah Poulson, 1790), 3.

36. Joel Barlow, *A Vision of Columbus* (Hartford, CT: Hudson and Goodwin, 1787), 9.247. On the manner in which yellow fever upset early republican hopes, see Eve Kornfeld, "Crisis in the Capital: The Cultural Significance of Philadelphia's Great Yellow Fever Epidemic," *Pennsylvania History* 51, no. 3 (July 1984), 189–205; and Finger, *Contagious City*, 103–151.

37. Benjamin Rush, *Observations on the Duties of a Physician* (Philadelphia: Prichard and Hall, 1789), 11.

38. William Barton, *Observations on the Progress of Population, and the Probabilities of the Duration of Human Life, in the United States of America* (Philadelphia: R. Aitken & Son, 1791), 2. For the response to Buffon, see Lee Alan Dugatkin, *Mr. Jefferson and the Giant Moose: Natural History in Early America* (Chicago: University of Chicago Press, 2009).

39. William Currie, *An Historical Account of the Climates and Diseases of the United States of America* (Philadelphia: Thomas Dobson, 1792), 193, 405.

40. Cathrall, *Medical Sketch of the Synochus Maligna*, 7.

41. Benjamin Rush, *An Enquiry into the Origin of the Late Epidemic Fever in Philadelphia* (Philadelphia: Mathew Carey, 1793), 9.

42. Rush, *Enquiry into the Origin*, 4.

43. Academy of Medicine of Philadelphia, Letter to Thomas Mifflin, December 1, 1797, in Rush, *Medical Inquiries and Observations* 5 (Philadelphia: Budd and Bartram, 1798), 44.

44. Mackrill, *History of the Yellow Fever*, 17–18.

45. William Currie, *Observations on the Causes and Cure of Remitting or Bilious Fevers* (Philadelphia: William T. Palmer, 1798), 41, 42.

46. Benjamin Rush, *An Account of the Bilious Remitting Yellow Fever, as It Appeared in the City of Philadelphia, in the Year 1793* (Philadelphia: Thomas Dobson, 1794), 95–97; Cathrall, *A Medical Sketch of the Synochus Maligna*, 6; Valentine Seaman, *An Account of the Epidemic Yellow Fever* (New York: Hopkins, Webb, 1796) 6.

47. This is an unduly controversial issue. Sheldon Watts, for example, has denied the notion of genetic immunity in Africans. See his "Yellow Fever Immunities in West Africa and the Americas in the Age of Slavery and beyond: A Reappraisal," *Journal of Social History* 34, no. 4 (Summer 2001), 955–967. The preponderance of the evidence suggests its likelihood. K. F. Kiple and V. H. Kiple, "Black Yellow Fever Immunities, Innate and Acquired, as Revealed in the American South," *Social Science and History* 1 (1977): 419–436; McNeill, *Mosquito Empires*, 44–45. For genomic evidence that shows yellow fever's age and place of origin, see Oyewale Tomori, "Yellow Fever: The Recurring Plague," *Critical Reviews in Clinical Laboratory Sciences* 41, no. 4 (2004), 397–398.

48. Seaman, *Account of the Epidemic Yellow Fever*, 20.

49. Currie, *Treatise on the Synochus Icteroides*, 12–13. Currie was partially wrong about his facts. Many people from Africa and the Mediterranean region with the sickle cell trait have an ability to resist the worst effects of malaria.

50. Ibid., 2.

51. Carey, *Short Account*, 16–20.

52. Matthew Livingstone David, *Brief Account of the Epidemical Fever which Lately Prevailed in the City of New York* (New York: Matthew L. Davis, 1795), 15–21.

53. College of Physicians, *Facts and Observations Relative to the Nature and Origin of the Pestilential Fever* (Philadelphia: Thomas Dobson, 1798), 22.

54. David Hume, *An Enquiry Concerning Human Understanding*, 7.21. There are many editions of the *Enquiry*. I have used the one in the Oxford Philosophical Texts series, ed. Tom L. Beauchamp (New York: Oxford University Press, 1998). For a cogent introduction to the history of theories of the mind, see Paul S. McDonald, *History of the Concept of the Mind* (Burlington, VT: Ashgate, 2003).

55. Benjamin Rush, *A Second Address to the Citizens of Philadelphia, Containing Additional Proofs of the Domestic Origin of the Malignant Bilious, or Yellow Fever* (Philadelphia: Budd and Bartram, 1799), iv.

56. Felix Pascalis Ouvière, *Medico-Chymical Dissertations on the Causes of the Epidemic Called Yellow Fever* (Philadelphia: Snowden & M'Corkle, 1796), 3.

57. James Delbourgo discusses the manifestations of common sense in the eighteenth century in "Common Sense, Useful Knowledge" 653–655.

58. Heiner F. Klemme, "Scepticism and Common Sense," in *The Cambridge Companion to the Scottish Enlightenment*, ed. Alexander Broadie (New York: Cambridge University Press, 2003), 118–119; M. A Stewart, "Religion and Rational Theology," in Broadie, *Cambridge Companion to the Scottish Enlightenment*, 60–78.

59. Delbourgo, "Common Sense, Useful Knowledge," 654. An excellent work on the common-sense philosophical roots of Scottish science is David B. Wilson, *Seeking Nature's Logic: Science in the Scottish Enlightenment* (University Park: Pennsylvania State University Press, 2009).

60. This perspective owes its existence to the perspicacious work of Mark Noll, *America's God: From Jonathan Edwards to Abraham Lincoln* (New York: Oxford University Press, 2002), esp. 93–113.

61. Virtually all of the major intellectual histories and historians agree on the importance of common-sense philosophy. Henry May, in *The Enlightenment in America* (New York: Oxford University Press, 1976), argues that common sense undergirded a distinctive phase of the American Enlightenment, which he calls the "Didactic Enlightenment," the dominant framework especially from 1800 to 1815. See also Richard Sher and Jeffrey Smitten, eds., *Scotland and America in the Age of the Enlightenment* (Edinburgh: Edinburgh University Press, 1990). For common sense in American universities and Protestant academies, see Douglas Sloan, *The Scottish Enlightenment and the American College Ideal* (New York: Teacher's College, Columbia University, 1971); and Mark Noll, *Princeton and the Republic: The Search for a Christian Enlightenment in the Era of Samuel Stanhope Smith* (Princeton, NJ: Princeton University Press, 1989).

62. Rush to James Kidd, May 13, 1794, in *Letters of Benjamin Rush* (Princeton, NJ: Princeton University Press, 1951), 2:748.

63. John Beale Davidge, *A Treatise on the Autumnal Endemial Epidemick of Tropical Climates, Vulgarly Called the Yellow Fever* (Baltimore: W. Pechin, 1798), 26.

64. Benjamin Rush, "Thoughts on Common Sense," in *Essays, Literary, Moral and Philosophical* (Philadelphia: Thomas and Samuel F. Bradford, 1798), 251, 255. Sloan dedicates a chapter to Benjamin Rush and his common sense in *Scottish Enlightenment*, 185–224.

65. Samuel Latham Mitchill, *The Present State of Learning in the College of New York* (New York: T. and J. Swords, 1794), title page. Mitchill quotes Bacon in the original Latin. The translation is my own.

66. Reid, for example, argued that the structures of languages bore witness to inherent mental powers—all languages distinguish between subject and object and all depict the operations of the mind in the active voice. Thomas Reid, *Essays on the Intellectual and Active Powers of Man* (Dublin: P. Byrne and J. Milliken, 1790); Alexander Broadie, "The Mind and Its Powers," in Broadie, *Cambridge Companion to the Scottish Enlightenment*.

67. Benjamin Rush, "On Education," unpublished essay, Library Company of Philadelphia Yi 2, 7400, F 34; Rush's "Lectures on the Mind" was later published as *Benjamin Rush's Lectures on the Mind*, ed. Eric T. Carlson et al. (Philadelphia: American Philosophical Society, 1981).

68. Benjamin Rush, *Medical Inquiries and Observations: Containing an Account of the Yellow fever, As it Appeared in Philadelphia in 1797* (Philadelphia: Thomas Dobson, 1798), v.

69. Rush, *Second Address to the Citizens*, 16–17.

70. Charles Caldwell, *A Semi-annual Oration, on the Origin of Pestilential Diseases* (Philadelphia: Thomas and Samuel Bradford, 1799), 18. The speech was given on December 17, 1798.

71. Currie, *Observations on the Causes*, 43.

72. Yellow fever certainly existed in endemic form in West Africa and West Indies ports, but not in the United States. Even the environmentalists knew it mostly as an epidemic disease. Henry Warren, *A Treatise Concerning the Malignant Fever in Barbados, and the Neighbouring Islands: With an Account of the Seasons there, from the year 1734 to 1738. In a letter to Dr. Mead* (London: Fletcher Gyles, 1741), 6, 1–9; John Lining, *A Description of the American Yellow Fever, which Prevailed at Charleston, in South Carolina, in the Year 1748* (Philadelphia: Thomas Dobson, 1799), 4–7.

73. Isaac Briggs, *Medical Repository* 6, no. 2 (April 1803), 170–171.

CHAPTER 2

1. David Hume, *An Enquiry Concerning Human Understanding*, 7.84.

2. Webster's biographers give only passing notice to his tome on disease history, treating it as a mere oddity—more a testimony to his eccentricity and erudition than an integral contribution to a pressing debate or as a unique window into early republican intellectual life. Horace Scudder, *Noah Webster* (Boston: Houghton,

Mifflin, 1890), 105, 161; Harry Warfel, *Noah Webster: Schoolmaster to America* (New York: Macmillan, 1936), 242–251; Harlow G. Unger, *Noah Webster: The Life and Times of an American Patriot* (New York: John Wiley and Sons, 1998), 209, 212. A detailed treatment of Webster's epidemiology appears in Charles-Edward Amory Winslow, *The Conquest of Epidemic Disease: A Chapter in the History of Ideas* (New York: Hafner, 1967), 208–232.

3. For key works on the influence of history on republican thought, see Pocock, *The Machiavellian Moment: Florentine Political Thought and the Atlantic Republican Tradition* (Princeton, NJ: Princeton University Press, 2003); Drew McCoy, *The Elusive Republic: Political Economy in Jeffersonian America* (Chapel Hill: University of North Carolina Press, 1980). For the special place of the classics in early republican thought, see Eran Shalev, *Rome Reborn on Western Shores: Historical Imagination and the Creation of the American Republic* (Charlottesville: University of Virginia Press, 2009); Caroline Winterer, *The Mirror of Antiquity: American Women and the Classical Tradition, 1750–1900* (Ithaca, NY: Cornell University Press, 2007); Carl J. Richard, *The Golden Age of the Classics in America: Greece, Rome, and the Antebellum United States* (Cambridge, MA: Harvard University Press, 2009). For works on formal history writing and history as a source of national identity, see Peter Messer, *Stories of Independence: Identity, Ideology, and Independence in Eighteenth-Century America* (DeKalb: Northern Illinois University Press, 2005); Eileen Ka-May Cheng, *The Plain and Noble Garb of Truth: Nationalism and Impartiality in American Historical Writing, 1784–1860* (Athens: University of Georgia Press, 2008). For American exceptionalism, see Dorothy Ross, "Historical Consciousness in Nineteenth-Century America," *American Historical Review* 89, no. 4 (October 1984), 909–928.

4. For a fascinating analysis of the use of history in the pursuit of disease and in medical thought more broadly, see Nancy Siraisi, *History, Medicine, and the Traditions of Renaissance Learning* (Ann Arbor: University of Michigan Press, 2007).

5. Benjamin Rush to Julia Rush, August 25, 1793, in *The Letters of Benjamin Rush*, ed. Lyman Butterfield (Princeton, NJ: Princeton University Press, 1951), 2:640.

6. Mathew Carey, *A Short Account of the Malignant Fever Lately Prevalent in Philadelphia*, 4th edition (Philadelphia: Mathew Carey, 1794), 24, 52. Carey took his accounts from *The History of the Great Plague in London, in the Year 1665 . . . By a Citizen, Who Lived the Whole Time in London*, printed in London in 1754. Though he did not know it at the time, the author was Daniel Defoe. For the history and impact of Defoe's work, see Margaret Healy, "Defoe's *Journal* and the English Plague Writing Tradition," *Literature and Medicine* 22, no. 1 (Spring 2003), 25–44.

7. "A Friend to the Public," *Federal Gazette*, September 27, 1793.

8. The same advertisement appeared frequently. See *Federal Gazette*, September 2, 1793. The *Federal Gazette* appeared every day for the duration of the epidemic. For an interesting take on the motivations of its editor, Andrew Brown, see Mark A. Smith, "Andrew Brown's 'Earnest Endeavor': The *Federal Gazette*'s Role in

Philadelphia's Yellow Fever Epidemic of 1793," *Pennsylvania Magazine of History and Biography* 120, no. 4 (October 1996), 321–342. The other newspaper to continue publication, though only sporadically, was *Dunlap's American Daily Advertiser.*

9. Richard Bayley, *An Account of the Epidemic Fever which Prevailed in the City of New York, during Part of the Summer and Fall of 1795* (New York: T. and J. Swords, 1796), 58.

10. Benjamin Rush, *An Enquiry into the Origin of the Late Epidemic Fever in Philadelphia* (Philadelphia: Mathew Carey, 1793), 7, 10.

11. Stubbins Ffirth, *An Inaugural Dissertation on Malignant Fever; with an Attempt to Prove Its Non-Contagious Nature, from Reason, Observation, and Experiment* (Philadelphia: B. Graves, 1804), 12–15.

12. John Beale Bordley, *Yellow Fever* (Baltimore: Charles Cist, 1794), 2, 8.

13. William Currie, *A Sketch of the Rise and Progress of Yellow Fever* (Philadelphia: Budd and Bartram, 1800), 40–42.

14. Matthew Livingstone Davis, *A Brief Account of the Epidemical Fever which lately Prevailed in the City of New York* (New York: M. L. Davis, 1795), 15.

15. Elihu Hubbard Smith, *The Diary of Elihu Hubbard Smith*, ed. James Cronin (Philadelphia: American Philosophical Society, 1973), September 7, 1795, 48.

16. Noah Webster, "Circular" (New York: Printed by author, 1795).

17. Louis W. McKeehan, *Yale Science: The First Hundred Years, 1701–1801* (New York: Henry Schuman, 1947), 21–22, 37–38, 47–55.

18. For biographical details, see K. Alan Snyder, *Defining Noah Webster: Mind and Morals in the Early Republic* (New York: University Press of America, 1990). For Webster's character and intellectual outlook, see Jill Lepore, *A is for American: Letters and Other Characters in the Newly United States* (New York: Vintage Books, 2002), 1–58.

19. Stephen Reynolds, "On the Fever in Montgomery County," in *A Collection of Papers on the Subject of Bilious Fevers*, ed. Noah Webster (New York: Hopkins, Webb, 1796), 195.

20. Eneas Monson, "An Account of the Fever in New Haven, in 1794," in Webster, *Collection*, 174–175.

21. Ibid., 175. A rod was a unit of measurement commonly in use in the eighteenth century, which denotes a length of about 5.5 yards.

22. Ibid., 176, 178.

23. Webster, *Collection*, 201.

24. For the history of the epidemic constitution, see Charles-Edward Amory Winslow, *The Conquest of Epidemic Disease: A Chapter in the History of Ideas* (New York: Hafner, 1967), 66, 106–110, 166–167, 172–174.

25. For biographical information about Miller, see William Huffington, *The Delaware Register* (Dover: S. Kimmey, 1839), 114–115; for the doings of the Friendly Club, see Bryan Waterman, *Republic of Intellect: The Friendly Club of New York and the Making of American Literature* (Baltimore: Johns Hopkins University, 2007).

26. Smith, *Diary of Elihu Hubbard Smith*, 201.

27. Edward Miller, Samuel Latham Mitchill, and Elihu Hubbard Smith, *An Address* (New York, 1796), 4.

28. Ibid., 2, 5.

29. Ibid., 1.

30. Smith, *Diary of Elihu Hubbard Smith*, Monday, November 14, 1796, 246.

31. Ibid., Thursday, July 7, 1796, 184, 246. Smith's essay on the Punic Wars was "On the Pestilential Diseases which, at different times, appeared in the Athenian, Carthaginian, and Roman Armies, in the neighbourhood of Syracuse," *Medical Repository* 2, no. 4 (May 1799), 367–382. For a history of *Medical Repository*, Patricia Kahn and Richard Kahn, "The *Medical Repository*—The First U.S. Medical Journal (1797–1824)," *New England Journal of Medicine* 337, no. 26 (December 1997), 1926–1930.

32. Elihu H. Smith, "The Plague of Athens," *Medical Repository* 1, no. 1 (August 1797), 15–16. Thucydides, *The History of the Peloponnesian War*, 2.48. The plague narrative is in 2.47–55.

33. Smith, "Plague of Athens," 26, 27.

34. Martha Slotten, "Elizabeth Graeme Ferguson: A Poet in the 'Athens of North America,'" *Pennsylvania Magazine of History and Biography* 108, no. 3 (July 1984), 259–288; Winterer, *Mirror of Antiquity*, 36; Kahn and Kahn, "*Medical Repository*—First U.S. Medical Journal," 1928. Richard, *Golden Age of the Classics in America*, 41–53, argues that early republicans focused on republican Rome at the expense of democratic Athens. This discussion, and indeed the entire thrust of this chapter, suggests an exception to that rule.

35. Webster's communications to Currie were collected and published by the editorial staff at the *Bulletin of the History of Medicine*. See Noah Webster, *Letters on Yellow Fever Addressed to Dr. William Currie*, Issue no. 9 of the *Supplements to the Bulletin of the History of Medicine*, ed. Henry Sigerist and with an introduction by Benjamin Spector (Baltimore: Johns Hopkins University Press, 1947), 22. In characteristic manner, Webster lamented that the confusion about the nature of contagions arose from the careless use of language. "It is to be wished, Sir, that words *infection* and *contagion*, had each a definite technical meaning. In ordinary language they are used as synonimous, and so are they explained in dictionaries."

36. Ibid., 30–31, 84.

37. Noah Webster, *Brief History of Epidemic and Pestilential Diseases* (Hartford, CT: Hudson and Goodwin, 1799), vii–viii.

38. Caroline Winterer provides an interesting commentary on women's readings of Rollin in *Mirror of Antiquity*, 26–29. For the use of travel narratives as pieces of empirical evidence in history and other inquiries, see Richard Olson, "The Human Sciences," in the *Cambridge History of Science*, vol. 4, *Eighteenth-Century Science* (New York: Cambridge University Press, 2003), 442–444. Another great treatment of the travel narratives is offered in P. J. Marshal and Glyndwr Williams, *The Great Map of Mankind: Perceptions of New Worlds in the Age of Enlightenment* (Cambridge, MA: Harvard University Press, 1982).

39. Webster, *Brief History*, 1:21–22.

40. Ibid., 1:22.

41. Ibid., 1:23.

42. Ibid., 1:23.

43. Ibid., 1:38.

44. Ibid., 2:10.

45. Ibid., 1:65–283.

46. Ibid., 2:9, 11.

47. Webster to Benjamin Rush, February 27, 1799, in *Letters on Yellow Fever*, 17.

48. Webster, *Brief History*, 2:15. Webster's implication of electricity underscored a widespread fascination with electricity in the early republic. See James Delbourgo, *A Most Amazing Scene of Wonders: Electricity and Enlightenment in Early America* (Cambridge, MA: Harvard University Press, 2006). His explanation also shows the persistence of marginalized practices including astrology. See Patricia Fara, "Marginalized Practices," in *Cambridge History of Science*, vol. 4, *Eighteenth-Century Science*, 486–507.

49. Noah Webster to Benjamin Rush, November 26, 1799, in *Letters of Noah Webster*, ed. Harry Warfel (New York: Library Publishers, 1953), 203–204.

50. Webster, *Brief History*, 2:78–134.

51. Ibid., 2:78, 79.

52. For Webster's character and nationalism, see Lepore, *A is for American*, 3–11. For European Enlightenment history, see Karen O'Brien, *Narratives of Enlightenment: Cosmopolitan History from Voltaire to Gibbon* (New York: Cambridge University Press, 1997. Ernst Breisach, *Historiography: Ancient, Medieval, and Modern* (Chicago: University of Chicago Press, 2007), 199–227.

53. See *Letters on Yellow Fever*, ed. Sigerist, 14.

54. "Review," *Medical Repository* 3, no. 3 (February 1800), 278.

55. Ibid., 279.

56. Ibid., 288.

57. *Medical Repository* 3, no. 4 (May 1800), 373.

58. Biographical information about Tytler can be found in James Ferguson, *Balloon Tytler* (London: Faber and Faber, 1972).

59. James Tytler, *Treatise on the Plague and Yellow Fever* (Salem, MA: Joshua Cushing for B.B. Macanulty, 1799), 22–24, 31–334, 464–470.

60. Ibid., 40–41, 47, 71.

61. This story is told in two places in the Bible: 2 Samuel 24; 1 Chronicles 21.

62. Tytler, *Treatise on the Plague and Yellow Fever*, 9. The story of Chryseis's abduction and the wrath of Apollo can be found in *Iliad*, Book 1.

63. Compared with Webster's, Tytler's history nicely illustrates an essential tension in eighteenth-century historiography. Like Webster's work and the dominant works of time, it universalized historical events—in this case, through the story of disease. Crucially, Webster recognized that diseases operated in cycles that appeared throughout history, and that they could be stopped or at least mitigated by

the progress of human enlightenment and the exercise of common sense. Tytler, rather, subordinated the world of humans and their decisions to a sacred history, with its more linear plotline, which placed God at both the beginning and the end. Carl Becker writes about this agreement and tension between the sacred and temporal, or enlightened, views of history in *The Heavenly City of Eighteenth-Century Philosophers*, 2nd ed. (New Haven, CT: Yale University Press, 2003 [1st ed. 1932]). Tytler and Webster (as well as the *philosophes* and theologians in Becker) wanted to get to the same end; only Tytler saw that end outside of history, whereas Webster saw it within history.

64. This is a broad picture of American historiographical development and is not meant to suggest that eighteenth-century historians utterly disavowed providentialism in history or that nineteenth-century historians disregarded secular processes in history. See Cheng, *Plain and Noble Garb of Truth*; Ross, "Historical Consciousness in Nineteenth-Century America."

65. *Medical Repository* 3, no. 4 (May 1800), 373–379.

66. William Chalwill, *A Dissertation on the Sources of Malignant Bilious, or Yellow Fever, and Means of Preventing It* (Philadelphia: Way and Groff, 1799), 8.

67. Phineas Jenks, *An Essay on the Analogy of the Asiatic and African Plague and the American Yellow Fever* (Philadelphia: Hugh Maxwell, 1804), 28.

68. Charles Caldwell, *A Semi-Annual Oration, on the Origin of Pestilential Diseases* (Philadelphia: Thomas and Samuel Bradford, 1799), 28, 29, 31.

69. William Currie, *Observations on the Causes and Cure of Remitting or Bilious Fevers* (Philadelphia: William T. Palmer, 1798), 11, 13.

70. Ibid., 15.

71. William Currie, *A Sketch of the Rise and Progress of the Yellow Fever, and of the Proceedings of the Board of Health, in Philadelphia, in the Year 1799* (Philadelphia: Budd and Bartram, 1800), 70–71.

72. Ibid., 68.

73. John Haygarth also challenged the efficacy of Elisha Perkins's metallic tractors—simple metallic rods supposed to cure a range of ailments by channeling the power of electricity. Citing experiments in which he tested patients who used Perkins's tractors against those who used imitation wooden tractors (which were incapable of conducting electricity), he concluded that the healing powers of tractors were produced by the minds of patients, not by the properties of the devices. See Haygarth, *On the Imagination as a Cause and as a Cure of Disease of the Body Exemplified by Fictitious Tractors and Epidemical Convulsions* (Bath: R. Cruttwell, 1800).

74. For the relationship, see Christopher Charles Booth, *John Haygarth, FRS (1740–1827)* (Philadelphia: American Philosophical Society, 2005), 125–127. Colin Chisholm, *An Essay on the Malignant Pestilential Fever Introduced into the West Indian Islands from Boullam on the Coast of Guinea* (Philadelphia: Thomas Dobson, 1799); *Observations and Facts, Tending to Prove that the Epidemic Existing at Philadelphia, New York &c Was the Same Fever Introduced by Infection Imported from the*

West India Islands (London: J. Mawman, 1801). Some of their lengthy correspondence was published contemporaneously. See, for example, Colin Chisholm, *Letter to John Haygarth from Colin Chisholm* (London: Joseph Mawman, 1809), a treatise that summarized Chisholm's thoughts about yellow fever.

75. The letter to the College of Physicians is featured in John Haygarth, *A Letter to Dr. Percival, on the Prevention of Infectious Fevers* (London: R. Cruttwell, 1801), 143–165. The quoted material is on p. 155.

76. My argument here about the way that historical inquiry undermined its own empirical, "scientific" pretensions bears much in common with Caroline Winterer, "Model Empire, Lost City: Ancient Carthage and the Science of Politics in Revolutionary America," *William and Mary Quarterly* 67, no. 1 (January 2010), 3–30.

77. Webster, *Brief History*, 2:248, 253. For a cogent discussion of the way that localism and yellow fever buttressed republican antiurbanism, see Finger, *Contagious City*, 153–162. For the way that historical thought as a whole reinforced antiurbanism, see McCoy, *Elusive Republic*.

78. Samuel Latham Mitchill in *An Account of the Malignant Fever, Lalely [sic] Prevalent in the City of New-York*, ed. James Hardie (New York: Hurtin and M'Farlane, 1799), 15, 16.

79. Bruno Latour, *The Pasteurization of France*, trans. Alan Sheridan and John Law (Cambridge, MA: Harvard University Press, 1988), 20–21, 32–33.

CHAPTER 3

1. John Stuart Mill, "The Subjection of Women," in *The Basic Writings of John Stuart Mill*, ed. Dale E. Miller (New York: Modern Library, 2002), 183.

2. Samuel Latham Mitchill, *Nomenclature of the New Chemistry* (New York: T. and J. Swords, 1794), 3, 4.

3. Marcellin Berthelot, *La Révolution Chimique: Lavoisier* (Paris, 1890).

4. Thomas Kuhn, *The Structure of Scientific Revolutions* (Chicago: University of Chicago Press, 1996), 111. Kuhn's picture of the nature of paradigms—their impact on human perception of nature—was partially drawn from research on cognition, behavior, and perception in the 1950s and 1960s. See pp. 111–135.

5. Information about Mitchill's early life and education can be gleaned from numerous sources. See Alan Aberbach, "Samuel Latham Mitchill: A Physician in the Early Days of the Republic," *Bulletin of the New York Academy of Medicine* 40, no. 7 (July 1964), 503; William Brock, *Scotus Americanus: A Survey of the Sources for Links between Scotland and America in the Eighteenth Century* (Edinburgh: Edinburgh University Press, 1982), 181; Graham Burnett, *Trying Leviathan: The Nineteenth-Century Court Case that Put the Whale on Trial and Challenged the Order of Nature* (Princeton, NJ: Princeton University Press, 2007), 44–52.

6. Samuel L. Mitchill, *Outline of the Doctrines in Natural History, Chemistry, and Economics* (New York: Childs and Swaine, 1792); Patricia Kahn and Richard

Kahn, "The *Medical Repository*—the First U.S. Medical Journal (1797–1824)," *New England Journal of Medicine* 337, no. 26 (December 25, 1997), 1926–1930.

7. John C. Greene, *American Science in the Age of Jefferson* (Ames: Iowa State University Press, 1984), 187.

8. Ibid., 158–187; Edgar Fahs Smith, *Chemistry in America: Chapters from the History of the Science in the United States* (New York: D. Appleton, 1914), especially 1–218; Sidney Edelstein, "The Chemical Revolution in the Pages of the Medical Repository," *Chymia* 5 (1959), 155–179.

9. Jan Golinski, "Chemistry," in *The Cambridge History of Science*, vol. 4, *Eighteenth-Century Science*, ed. Roy Porter (Cambridge: Cambridge University Press, 2003), 375–396. Robert Siegfried, "The Chemical Revolution in the History of Chemistry," *Osiris*, 2nd ser., 4 (January 1988), 34–50. Siegfried argues that the truly revolutionary aspect of the chemical revolution for its contemporaries was its revelations about the elemental construction of nature.

10. Anders Lundgren, "The New Chemistry in Sweden: The Debate that Wasn't," *Osiris*, 2nd ser., 4 (1988), 146–168; H.A.M. Snelders, "The New Chemistry in the Netherlands," *Osiris*, 2nd ser., 4 (1988), 121–145; Ramón Gago, "The New Chemistry in Spain," *Osiris*, 2nd ser., 4 (1988), 169–192.

11. For the spread of the new chemistry in the United States, see Edgar F. Smith, *Chemistry in America*.

12. Felix Pascalis Ouvière, *Annual Oration Delivered before the Chemical Society, January 31, 1801* (Philadelphia: John Bioren, 1802), 5–6, found in the Malloch Rare Book Room, New York Academy of Medicine.

13. Joseph Browne, *Treatise on the Yellow Fever* (New York: from the Office of the Argus, 1798), 6–7.

14. Thomas Hankins, *Science and the Enlightenment* (New York: Cambridge University Press, 1985), 81–84.

15. Browne, *Treatise on the Yellow Fever*, 6–7. Benjamin De Witt, "A Chemico-Medical Essay to the Explain the Operation of Oxigene, or the Base of Vital Air on the Human Body" (Philadelphia: William Woodward, 1797), 35, in *Bound Pamphlets* 331 at the New York Academy of Medicine.

16. Hankins, *Science and the Enlightenment*, 85–92.

17. A clear contemporary account of fermentation comes from Antoine Fourcroy, *Chemical Philosophy; or, the Established Bases of Modern Chemistry*, trans. W. Desmond (London: H. D. Symonds, 1807). For some thoughts on fermentation in the eighteenth-century imagination, see Alain Corbin, *The Foul and the Fragrant: Odor and the French Social Imagination*, trans. Miriam Kochan (Cambridge: Harvard University Press, 1986).

18. Samuel Latham Mitchill, *Nomenclature of the New Chemistry* (New York: T. and J. Swords, 1794), 11.

19. Antoine-Laurent Lavoisier, *Traité Élémentaire de Chimie* (Paris: Cuchet, 1789), xviii. For a thoughtful consideration of French empiricism in the late eighteenth century, see Jessica Riskin, *Science in the Age of Sensibility: The Sentimental*

Empiricists of the French Enlightenment (Chicago: University of Chicago Press, 2002), esp. 227–282.

20. Quoted in Arthur Donovan, "Scottish Responses to the New Chemistry of Lavoisier," *Studies in Eighteenth Century Culture* 9 (1979), 239; Donovan, *Philosophical Chemistry in the Scottish Enlightenment* (Edinburgh: Edinburgh University Press, 1984); for Black's theoretical chemistry, which he taught from 1766 until shortly before his death in 1799, see David B. Wilson, *Seeking Nature's Logic: Science in the Scottish Enlightenment* (University Park: Pennsylvania State University Press, 2009), 133–169. Wilson shows that a common-sense tradition pervaded Scottish science in the eighteenth century.

21. Samuel Latham Mitchill, "Answer to the Two Letters from Dr. Priestley," *Medical Repository* 2, no. 1 (July 1798), 51–54. This confidence was rooted in a higher level of instrumental precision. See Jan Golinski, "Precision Instruments and the Demonstrative Order of Proof in Lavoisier's Chemistry," *Osiris*, 2nd ser., 9 (1994), 30–47.

22. His suggestions never stuck. Samuel Latham Mitchill, *Explanation of the Synopsis of Chemical Nomenclature and Arrangement: Containing Several Important Alterations of the Plan Originally Reported by the French Academicians* (New York: T. and J. Swords, 1801).

23. Of course, Priestley also asserted that he was not ready to surrender like Wurmser. Joseph Priestley, *Medical Repository* 1, no. 2 (November 1797), 521–522. Priestley's complaint was a decisive one. Golinski has shown how British opinion about Lavoisier's chemistry turned on their objections to the Lavoisierians' overbearing attitudes. See Jan Golinski, *Science as Public Culture: Chemistry and Enlightenment in Britain, 1760–1820* (New York: Cambridge University Press, 1992), 129–152; Golinski, "The Chemical Revolution and the Politics of Language," *The Eighteenth Century* 33, no. 3 (Fall 1992), 238–251.

24. For American reactions to the French Revolution, see Rachel Hope Cleves, *The Reign of Terror in America: Visions of Violence from Anti-Jacobinism to Antislavery* (New York: Cambridge University Press, 2009), 1–19, 58–103; David Brion Davis, *Revolutions: Reflections on American Equality and Foreign Liberations* (Cambridge, MA: Harvard University Press, 1990), 27–54. The first groups to turn against the revolution were the political and religious conservatives of the northeast, then those in the middle states, and finally the southerners, who deemed that the examples of actual slave rebellions were sufficient causes to denounce revolutionary bloodshed in general. Fear of the French Revolution, of the secret plots that reduced it, crystallized in the Bavarian Illuminati conspiracy, the best account of which is Vernon Stauffer, *New England and the Bavarian Illuminati* (New York: Columbia University Press, 1918).

25. In Scotland, the other major bastion of common-sense learning, John Robison—successor to Joseph Black as professor of chemistry at the University of Edinburgh and an opponent of the French Revolution and French philosophies—turned against French chemistry. When he edited and published the lectures of

Joseph Black, who had adopted the Lavoisierian system in 1790, Robison altered the content in an attempt "to portray his master as an ally against the French." See Donovan, "Scottish Responses to the New Chemistry of Lavoisier," 246. Robison was the author of *Proofs of a Conspiracy against All the Religions and Governments of Europe, Carried on in the Secret Meetings of Freemasons, Illuminati and Reading Societies* (1797), the text that sparked the Bavarian Illuminati conspiracy. In the anxious 1790s, the line between French chemistry and the French Revolution was often blurred. Consider another example. In his denunciation of the French Revolution, Edmund Burke used chemical metaphors to describe the spread of revolutionary enthusiasm: "The wild *gas*, the fixed air, is plainly broke loose: but we ought to suspend our judgment until the effervescence is a little subsided, till the liquor is cleared, and until we see something deeper than the agitation of a troubled and frothy surface." See Edmund Burke, *Reflections on the Revolution in France* (Oxford: Oxford University Press, 1993 [1st ed. 1790]), 8.

26. Henry Guerlac, "Chemistry as a Branch of Physics: Laplace's Collaboration with Lavoisier," in *Historical Studies in the Physical Sciences*, vol. 7, ed. Russell McCormmach (Princeton, NJ: Princeton University Press, 1976), 193–276.

27. W. R. Albury, "The Logic of Condillac and the Structure of French Chemical and Biological Theory, 1780–1801," PhD diss., Johns Hopkins University, 1972.

28. Lavoisier, *Traité Élémentaire*, xvi.

29. Benjamin Rush, *Medical Inquiries and Observations: Containing an Account of the Yellow Fever, As it Appeared in Philadelphia in 1797* (Philadelphia: Budd and Bartram, 1798), v.

30. Samuel Latham Mitchill, *Medical Repository* 4, no. 1, 389–390.

31. Benjamin Rush, *The Letters of Benjamin Rush*, ed. Lyman Butterfield (Princeton, NJ: Princeton University Press), 1:28, 41.

32. Benjamin Rush, *Medical Inquiries and Observations, Containing an Account of the Bilious and Remitting and Intermitting Yellow Fever* (Philadelphia: Thomas Dobson, 1796), 75–77.

33. Benjamin Rush, *Benjamin Rush's Lectures on the Mind*, ed. Eric T. Carlson, Jeffrey L. Wollock, and Patricia S. Noel (Philadelphia: American Philosophical Society, 1981), 181.

34. Alexander Hosack, *An Inaugural Essay on the Yellow Fever* (New York: T. and J. Swords, 1797), 9–11.

35. Unsurprisingly, the rage of eudiometry never caught on in the United States. Golinski, *Science as Public Culture*, 117–128.

36. Rush, *Medical Inquiries and Observations, Containing an Account of . . . Yellow Fever*, 76, 77.

37. William Currie, *Observations on the Causes and Cure of Remitting or Bilious Fevers* (Philadelphia: William T. Palmer, 1798), 15, 40.

38. Despite his admiration for French chemists—"the names of Hales, Black, Macbride, Priestly, Cavendish, Lavoisier, Fourcroy, Berthollet, Chaptal and De Morveau will not be forgotten while science is recorded," he wrote—Browne

disliked the terms that Lavoisier and his collaborators selected. Browne, *Treatise on the Yellow Fever*, 6–12.

39. Ibid., 15.

40. Ibid., 28–29.

41. My discussion of romanticism and science has been drawn from the essays in Andrew Cunningham and Nicholas Jardine, eds., *Romanticism and the Sciences* (New York: Cambridge University Press, 1990); Robert J. Richards, *The Romantic Conception of Life: Science and Philosophy in the Age of Goethe* (Chicago: University of Chicago Press, 2002). The term "cold philosophy" comes from Keats's "Lamia," 2.230—"Do not all charms fly / at the mere touch of cold philosophy?" Goethe's complaint about the heartlessness of knowledge "extorted with levers and screws" comes in his *Faust*, 1.674–675. For his involvement in romantic science, Richards, *Romantic Conception of Life*.

42. The standard account has it that romanticism in American science begins with Ralph Waldo Emerson in the 1830s, especially with the publication of his essay "Nature" (1836). Laura Dassow Walls, *Emerson's Life in Science: The Culture of Truth* (Ithaca, NY: Cornell University Press, 2003); David M. Robinson, *Natural Life: Thoreau's Worldly Transcendentalism* (Ithaca, NY: Cornell University Press, 2004).

43. For the natural historical and scientific roots of American romanticism, see Richard W. Judd, *The Untilled Garden: Natural History and the Spirit of Conservation in America, 1740–1840* (New York: Cambridge University Press, 2009); Aaron Sachs, *The Humboldt Current: Nineteenth Century Exploration and the Roots of American Environmentalism* (New York: Penguin, 2007). Both see a romantic impulse taking root before Emerson's time and finding expression in Americans' wonder in the face of the boundless and varied landscape.

44. Elizabeth Ann Williams, *A Cultural History of Medical Vitalism in Enlightenment Montpelier* (New York: Ashgate, 2003).

45. Felix Pascalis Ouvière, *Medico-Chymical Dissertations on the Causes of the Epidemic Called Yellow Fever* (Philadelphia: Snowden and M'Corkle, 1796), 3, 7–8, 11.

46. Ibid., 11–23. Ouvière's notion laid itself open to the classic critique of the contagionists, the causal inconsistency, because again, why on earth would yellow fever occur in the caloric-ridden environs of the U.S. cities and not all caloric-ridden places? Ouvière answered that it occurred because Americans, with their "excellent pasturage," ate too much beef, which made the body vulnerable to caloric. Hence, people in other hot areas did not get yellow fever.

47. Ibid., 3. The passage is derived from Ovid, *Metamorphoses*, 2.218–219. Ouvière added the word "*corpus*" to his treatise.

48. Ouvière, *Annual Oration*, 8–9.

49. Ibid., 12.

50. 80° Reaumur marked the boiling point of water, the equivalent of 212° Fahrenheit or 100° Celsius.

51. Samuel Latham Mitchill, *Remarks on the Gaseous Oxyd of Azote or of Nitrogene* (New York: T. and J. Swords, Printers of the Faculty of Physic of Columbia College, 1795), Preface.

52. Joseph Priestley, *Experiments and Observations on Different Kinds of Air* (Birmingham: Thomas Pearson, 1790), 2.54–55. Priestley claimed of it that it was an "air in which a candle burns quite naturally and freely, and which is yet in the highest degree noxious to animals, insomuch as they die the moment they are put into it."

53. Lavoisier, *Traité Élémentaire*, 82.

54. See the Department of Health and Human Services website at http://www .atsdr.cdc.gov/mmg/mmg.asp?id=394&tid=69, accessed on August 20, 2014. The other potential candidate for the gaseous oxyd of azote is the compound nitrogen dioxide (NO_2). This appears unlikely, however, because nitrogen dioxide exists as a fluid at any temperature above 70° F, and Mitchill, Lavoisier, and Priestley all knew it only as a gas.

55. Mitchill, *Remarks on the Gaseous Oxyd of Azote*, 9–11.

56. Ibid., 13–14.

57. Ibid., 14.

58. *Aedes aegypti* are most active when the temperatures are around 82–83° Fahrenheit. They will not bite when the temperature is below 77° or over 105°. Rickard Christophers, *Aedes Aegypti (L.), The Yellow Fever Mosquito: Its Life History, Bionomics and Structure* (Cambridge: Cambridge University Press, 1960), 474–475.

59. Mitchill endorses the use of lime and other alkalines in a number of essays. See Samuel Latham Mitchill, "Concerning the Use of Alkaline Remedies in Fevers, and the Analogy between Septic Acids and Other Poisons," *New York Magazine, or Literary Repository* (April 1797); and *Medical Repository* 4, no. 3 (January 1801), 297–326.

60. Samuel L. Mitchill, *Hints toward Promoting the Health and Cleanliness of New York City* (New York: T. and J. Swords, 1802), 5.

61. Edward Miller, *Medical Repository* 2, no. 4 (May 1799), 409.

62. Felix Pascalis Ouvière, *Medical Repository* 3, no. 4 (May 1800), 346.

63. Joseph Priestley, quoted by Samuel Latham Mitchill, *Medical Repository* 3, no. 3 (February 1800), 307.

64. See, for example, J.E.R. Birch, *Medical Repository* 3, no. 3 (February 1800), 308; Jeremiah Barker, *Medical Repository* 6, no. 1 (January 1803), 18–24; Isaac Briggs, *Medical Repository* 6, no. 2 (April 1803), 168–172.

65. Samuel Latham Mitchill, with Edward Miller and Elihu Hubbard Smith, *An Address* (New York, 1796), 1.

66. John W. Francis, *Reminiscences of the Late Samuel Latham Mitchill* (New York, 1866), 30.

67. Mitchill, *Remarks on the Gaseous Oxyd of Azote*, 42–43.

68. Isaac Cathrall, *A Medical Sketch of the Synochus Maligna, or Malignant Contagious Fever* (Philadelphia: Thomas Dobson, 1794).

69. Ibid., iv. Biographical information about Cathrall comes from James Thacher, *American Medical Biography: Or, Memoirs of Eminent Physicians Who Have Flourished in America* (Boston: Richardson and Lord, and Cottons and Barnard, 1828), 214–217. For hostility and resistance to autopsy, see Michael Sappol, *A Traffic of Bodies: Anatomy and Embodied in Nineteenth-Century America* (Princeton, NJ: Princeton University Press, 2002).

70. For anatomy as an experimental discipline, see Andrew Cunningham, *An Anatomist Anatomis'd: An Experimental Discipline in Enlightenment Europe* (Burlington, VT: Ashgate, 2010).

71. Rush to John Morgan, October 22, 1768, in *Letters of Benjamin Rush*, 1:66.

72. Isaac Cathrall, *Memoir on the Analysis of the Black Vomit* (Philadelphia: R. Folwell, 1800), 19. Pouppé Desportes, *Histoire des Maladies de Saint-Domingue* (Paris: Chez Lejay, 1770), 202–203. The work was based on Desportes's experience in Saint-Domingue during the first half of the eighteenth century. He died in 1748.

73. Cathrall, *Analysis of the Black Vomit*, 17–18.

74. Ibid., 19–22.

75. Anonymous, *Experiments on the Black Vomit* (Philadelphia: Bartholomew Graves), 54.

76. Ibid., 54.

77. Ibid., 56–57.

78. James Delbourgo, *A Most Amazing Scene of Wonders: Electricity and Enlightenment in Early America* (Cambridge, MA: Harvard University Press, 2006). Delbourgo argues that the bodily experiences (wonder, pain, surprise) of electrical experiments were themselves central to the science.

79. Samuel Brown, *An Inaugural Dissertation on the Bilious Malignant Fever* (Boston: Manning and Loring, 1797), 8.

80. Edgar Fahs Smith, *Chemistry in America*, 152–351.

81. Samuel Latham Mitchill, "Doctrine of Septon," *Medical Repository* 1, no. 2 (November 1797), 189–190.

CHAPTER 4

1. Quoted in John Adams, *The Works of John Adams, Second President of the United States*, ed. Charles Francis Adams (Boston: Little, Brown, 1854), 128.

2. Mitchill's public reputation would be fixed by 1818. See Graham Burnett, *Trying Leviathan: The Nineteenth-Century Court Case that Put the Whale on Trial and Challenged the Order of Nature* (Princeton, NJ: Princeton University Press, 2007), 44–52; John W. Francis, *Reminiscences of the Late Samuel Latham Mitchill* (New York, 1866), esp. 18–19, where he notes that Mitchill "suffered the stings of satire long and deeply."

3. Len Travers, *Celebrating the Fourth: Independence Day and the Rites of Nationalism in the Early Republic* (Amherst: University of Massachusetts Press, 1997);

David Waldstreicher, *In The Midst of Perpetual Fetes: The Making of American Nationalism, 1776–1820* (Chapel Hill: University of North Carolina Press, 1997).

4. Samuel Latham Mitchill, "An Address to the Citizens of New-York, Who Assembled in the Brick Presbyterian Church, to Celebrate the Twenty-third Anniversary of American Independence" (New York, 1800), 2.

5. Daniel Dennett has argued that this last trace of Aristotelianism encountered its first real philosophical challenge from the work of Charles Darwin, who demonstrated that a thing, a living thing especially, might exist for no apparent purpose at all, but only because it had survived through natural selection. Daniel Dennett, *Darwin's Dangerous Idea: Evolution and the Meanings of Life* (New York: Simon and Schuster, 199). Darwin had not disproved the Creator, nor even tried, but he had demonstrated that the cause of a thing, such as, say, the cause of the specific qualities of a cat, could not always be grasped through a consideration of the purpose of its creator, since the cat did not gain its qualities from a creator, but from the accumulation of the variations and mutations that resulted in the survival of the cat and its evolutionary ancestors.

6. Admittedly, natural theology means different things to different people. At its core, it is a way of investigating God through nature, without the aid of scripture (it is also sometimes called rational theology). The term itself only goes back to 1801, with William Paley's *Natural Theology*. Paley, however, used the evidence of nature to argue for the existence of the Christian God and to confirm the scriptures. The investigators used the evidence of design for the same purpose, though they never used the term "natural theology." Thus my own use of the term is meant to reflect their understanding of a type of evidence and its applications.

7. My own picture of the commercial orientation of the localists as well as most others owes much to the argument pioneered by Drew McCoy, *The Elusive Republic: Political Economy in Jeffersonian America* (Chapel Hill: University of North Carolina Press, 1980).

8. As far as I know, this reconciliation of divine and natural causes has gone unacknowledged by historians, though Charles Rosenberg was close. He showed that Americans of the 1830s saw cholera as a divine retribution for sin. The means through which God operated, however, were predisposing causes, especially intemperance, which prepared the body to accept disease. "In this doctrine of predisposing causes, the needs and attitudes of an awakening science found practical reconciliation with the ancient, and reassuring, idea of sin as a cause of disease." See Rosenberg, *Cholera Years* (Chicago: University of Chicago Press, 1962), 40, 40–54. The fever investigators, though, believed that the very material cause of yellow fever emerged from sinful behavior, consistent with God's will, of course. Though seemingly a minute difference, it did have drastic theological meanings. Rosenberg's realization explains why some people got the disease and some did not, but it could not grasp the purpose for the material cause of the disease (i.e., why it even existed in the created world).

9. The advent and subsequent spread of monotheistic religions in the ancient Near East exacerbated these problems. Polytheists could always attribute evil to wiles of the gods, who would sometimes target humans with their powers, but monotheists had no such recourse. See James P. Allen et al., "Theology, Theodicy, Philosophy," in *Religions of the Ancient World: A Guide*, ed. Sarah Iles Johnston (Cambridge, MA: Belknap Press of Harvard University Press, 2004), 531–546.

10. Leibniz gives this argument, famously ridiculed in Voltaire's *Candide*, in his *Theodicy* (1709).

11. Genesis 12:17; Exodus 8–12. Charles-Edward Amory Winslow, *The Conquest of Epidemic Disease* (Madison: University of Wisconsin Press, 1980), 35–39, offers an excellent introduction and compendia of such biblical verses.

12. Leviticus 26:14–46; Jeremiah 14:11; and for a more general overview see Kings 1 and 2, and Samuel 1 and 2.

13. On the union of colonial science and religion: Sarah Rivett's *The Science of the Soul in Colonial New England* (Princeton, NJ: Princeton University Press, 2011); for a broader introduction to the overlaps between science and religion, a good, though older, work is David C. Lindberg and Ronald L. Numbers, eds., *God and Nature: Historical Essays on the Encounter between Christianity and Science* (Berkeley: University of California Press, 1986). For an excellent example of the confusion of divine and natural explanations for disease specifically, see Anthony Kaldellis, "Literature of Plague and Anxieties of Piety in Sixth-Century Byzantium," in *Piety and Plague: From Byzantium to the Baroque*, eds. Franco Mormando and Thomas Worcester (Dexter, MO: Truman State University Press, 2007), 1–22.

14. Boccaccio, *Decameron*, trans. George Henry McWilliam (New York: Penguin, 2003), introduction. See Anna Montgomery Campbell, *The Black Death and Men of Learning* (New York: Columbia University Press, 1931), 37–49.

15. Nathan Hatch, *The Democratization of American Christianity* (New Haven, CT: Yale University Press, 1989); Mark Noll, *America's God: From Jonathan Edwards to Abraham Lincoln* (New York: Oxford University Press, 2002), esp. 161–208.

16. Noll, *America's God*, 171.

17. Anonymous, *An Earnest Call, Occasioned by the Alarming Pestilential Contagion* (Philadelphia: Jones, Hoff, and Derrick, 1793), 7.

18. For Green's involvement in Princeton science, see Mark Noll, *Princeton and the Republic: The Search for a Christian Enlightenment in the Era of Samuel Stanhope Smith* (Princeton, NJ: Princeton University Press, 1989), 272–292.

19. Ashbel Green, *A Pastoral Letter from a Minister in the Country, to Those of His Flock Who Remained in Philadelphia during the Pestilence of 1798* (Philadelphia: John Ormand, 1799), 5, 8. On millennialism in the early republic, Ruth Bloch, *Visionary Republic: Millennial Themes in American Thought, 1756–1800* (New York: Cambridge University Press, 1988), esp. 119–232.

20. Martin Pernick, "Politics, Parties, and Pestilence: Epidemic Yellow Fever in Philadelphia and the Rise of the First Party System," in *A Melancholy Scene of*

Devastation: The Public Response to the 1793 Philadelphia Yellow Fever Epidemic (Canton, MA: Science History Publications, 1997), 126.

21. Gary Nash, *Forging Freedom: The Formation of Philadelphia's Black Community, 1720–1835* (Cambridge, MA: Harvard University Press, 1991), 123.

22. Mathew Carey, *A Short Account of the Malignant Fever Lately Prevalent in Philadelphia* (Philadelphia: Mathew Carey, 1793), 11–12.

23. Mathew Carey, *Observations on Dr. Rush's Enquiry into the Origin of the Late Epidemic Fever in Philadelphia* (Philadelphia: Mathew Carey, 1793).

24. Benjamin Rush to Julia Rush, in *The Letters of Benjamin Rush*, ed. Lyman Butterfield (Princeton, NJ: Princeton University Press, 1951), 2:642–643, 656, 657, 663. The story of Shadrach, Meshach, and Abednego comes from the book of Daniel 3:16–27.

25. Rush uses this phrase at least twice. Ibid., 2:727, 733.

26. Rush to Julia, ibid., 2:689, 727.

27. Rush to Julia, August 29, 1793, ibid., 2:645. Rush's escape from harm was improbable. Mathew Carey recalled that at least ten physicians died and "hardly one" escaped infection. See Carey, *Short Account*, 72.

28. Rush to Julia, ibid., 2:645, 679.

29. Rush to Elias Boudinot, September 25, 1793, ibid., 2:681.

30. "Review," *Medical Repository* 3, no. 4 (May 1800), 374, 376.

31. The historiography usually depicts the demise of spontaneous generation coming from the experimental work of Pasteur and being epitomized by the famous argument between Pasteur and Felix Pouchet in the French Academy of Sciences. See Maurice Crosland, *Science under Control: The French Academy of Sciences 1795–1914* (Cambridge: Cambridge University Press, 2002), 118–119. Clara Pinto-Correia, however, finds signs of its early "fall from grace" coming in the middle of the seventeenth century, notwithstanding its endorsement from William Harvey in *Exercitationes de generatione animalium* (1651). See *The Ovary of Eve: Egg and Sperm and Preformation* (Chicago: University of Chicago Press, 1997), 3.

32. *Medical Repository* 3, no. 4 (May 1800), 393.

33. Joseph Priestley, *Medical Repository* 5, no. 1 (January 1802), 34–35.

34. Noah Webster, *Medical Repository* 5, no. 1 (January 1802), 30.

35. Ibid., 30–31.

36. E. Brooks Holifield, *Theology in America: Christian Thought from the Age of the Puritans to the Civil War* (New Haven, CT: Yale University Press, 2003), 159–196, discusses the threat of deism and the theological responses. Jon Butler also discusses the problem of deism in *Awash in A Sea of Faith: Christianizing the American People* (Cambridge, MA: Harvard University Press, 1992), 212–220.

37. Benjamin Rush to Ashbel Green, December 9, 1802, in *Letters of Benjamin Rush*, 2:853.

38. Herbert M. Morais, *Deism in Eighteenth-Century America* (New York: Russell and Russell, 1960), 120.

39. Ethan Allen, *Reason the Only Oracle of Man, or a Compendious System of Natural Religion* (Bennington: Haswell and Russell, 1784), 235; Elihu Palmer, *Principles of Nature; or, a Development of the Moral Causes of Happiness and Misery among the Human Species* (New York, 1801), 65, 69; Holifield discusses these arguments in *Theology in America*, 168.

40. The essential parameters of the debate resemble the dispute between Gottfried Wilhelm Leibniz, who argued that God existed outside the created world whose laws did not require maintenance, and Samuel Clarke (speaking for Isaac Newton), whose omnipresent God pervaded creation and maintained its laws. See G. W. Leibniz and Samuel Clarke, *Correspondence*, ed. Roger Ariew (Indianapolis: Hackett Publishing, 2000).

41. Samuel Latham Mitchill, *Medical Repository* 3, no. 4 (December 1800), 377.

42. Ibid., 377–378.

43. Samuel Latham Mitchill in *An Account of the Malignant Fever*, ed. James Hardie (New York: Hurtin and M'Farlane, 1799), 16.

44. Benjamin Rush, *Observations upon the Origin of the Malignant Bilious, or Yellow Fever* (Philadelphia: Budd and Bartram, 1799), 26. Some historians have argued that biblical injunctions against dirty things reflected basic disease-avoidance behaviors. William McNeill notes, for example, that Jewish and Muslim prohibitions against pork had a salutary effect because pigs carry many diseases. See *Plagues and Peoples* (New York: Anchor Books, 1998 [1st ed. 1976]), 64. Mary Douglas suggests another explanation. According to her, the things that we deem dirty are those ambiguous things that fall outside of the conceptual categories that organize and define the world. Pigs were prohibited in the Bible because they were cloven-hoofed and yet did chew their own cud, a disorienting anomaly and reason enough for its banishment to the realm of the unclean. Mary Douglas, *Purity and Danger: An Analysis of the Concepts of Pollution and Taboo* (New York: Routledge, 2002 [1st ed. 1966]), 51–71.

45. Noah Webster, *Brief History of Epidemic and Pestilential Diseases* (Hartford, CT: Hudson and Goodwin, 1799), 27.

46. Rush, *Observations upon the Origin of the Malignant Bilious, or Yellow Fever*, 10–11.

47. Webster, *Brief History*, 27.

48. Mary Douglas, *Purity and Danger*; Simon Finger highlights the connections between public health and republicanism in the early Philadelphia in *The Contagious City: The Politics of Public Health in Early Philadelphia* (Ithaca, NY: Cornell University Press, 2012), 120–162; the theme of public health as social control is prominent in Andrew Aisenberg, *Contagion: Disease, Government, and the "Social Question" in Nineteenth-Century France* (Stanford, CA: Stanford University Press, 1999).

49. Richard L. Bushman and Claudia L. Bushman, "The Early History of Cleanliness in America," *Journal of American History* 74, no. 4 (March 1988), 1214.

50. Kathleen M. Brown, *Foul Bodies: Cleanliness in Early America* (New Haven, CT: Yale University Press, 2009), 195–205; Dell Upton emphasizes the changing landscapes of urban cities and their effects on urban design in *Another City: Urban Life and Urban Spaces in the New American Republic* (New Haven, CT: Yale University Press, 2008).

51. Rush, *Letters of Benjamin Rush*, 2:653.

52. Charles Caldwell, *A Semi-Annual Oration, on the Origin of Pestilential Diseases, Delivered before the Academy of Medicine of Philadelphia* (Philadelphia: Thomas and Samuel Bradford, 1799), 36.

53. Indeed, by the time of the epidemics, Christian and republican principles had come together for most Americans in what Mark Noll describes as the "republican synthesis"—a term denoting the "fundamental compatibility between orthodox Protestant religion and republican principles of government"—such that to speak of one was necessarily to speak of the other. See Mark Noll, *America's God*, the quote is from 54; for the republican synthesis, see 53–92.

54. Samuel Latham Mitchill, *Medical Repository* 3, no. 4 (May 1800), 377.

55. Thomas Jefferson to Benjamin Rush, September 23, 1800, in *The Papers of Thomas Jefferson* (Princeton, NJ: Princeton University Press, 2005), 32:458–459.

56. Benjamin Rush to Thomas Jefferson, October 6, 1800, in *Letters of Benjamin Rush*, 2:824.

57. One can see the reformist spirit strongly in Richard Bayley, *An Account of the Epidemic Fever which Prevailed in the City of New York, during Part of the Summer and Fall of 1795* (New York: T. and J. Swords, 1796). Historians have written extensively about yellow fever's impacts on the public, but they have either missed or downplayed its religious dimension. See John Duffy, *The Sanitarians: A History of American Public Health* (Urbana-Champaign: University of Illinois Press, 1990), 38–48; Martin Melosi, *The Sanitary City: Urban Infrastructure in America from Colonial Times to the Present* (Baltimore: Johns Hopkins University Press, 2000), 30–31; Finger, *The Contagious City*, 120–151.

58. Noah Webster, *Collection of Papers on the Subject of Bilious Fevers* (New York: Hopkins, Webb, 1796), 244. Webster's recommendations suggest his debt to Hebraic thought, a subject in which scholars have shown recent interest. See, for example, Nathan R. Perl-Rosenthal, "The 'divine right of republics': Hebraic Republicanism and the Debate over Kingless Government in Revolutionary America," *William and Mary Quarterly* 66, no. 3 (July 2009), 535–564. Eric Nelsen, *The Hebrew Republic: Jewish Sources and the Transformation of European Political Thought* (Cambridge, MA: Harvard University Press, 2011).

59. Rush, *Observations upon the Origin*, 26.

60. Ibid., 26.

61. Mitchill in Hardie, ed., *Account of the Malignant Fever*, 16–17.

62. Caldwell, *A Semi-Annual Oration on the Origin of Pestilential Diseases*, 16.

63. *Medical Repository* 3, no. 4 (May 1800), 378.

64. For early republican political economy and its capitalist leanings, see Drew McCoy, *The Elusive Republic: Political Economy in Jeffersonian America* (Chapel Hill: University of North Carolina Press, 1980); Joyce Appleby, *Capitalism and a New Social Order: The Republican Vision of the 1790s* (New York: NYU Press, 1984). Their view of political economy fit closely with that outlined in Adam Smith's influential *Wealth of Nations* (1776). The role of providence and natural theology (the evidence of design) in Adam Smith's theory of economics has often been discussed. See Jacob Viner, *The Role of Providence in the Social Order* (Philadelphia: American Philosophical Society, 1972), 55–85, esp. 79–85; Lisa Hill, "The Hidden Theology of Adam Smith," *Journal of Economic Thought* 8, no. 1 (Spring 2001), 1–29. McCoy also discusses the seeming naturalness of commerce in the minds of early republicans in *Elusive Republic.*

65. Benjamin Rush, *A Plan for the Establishment of Public Schools and the Diffusion of Knowledge in Pennsylvania; to which Are Added Thoughts upon the Mode of Education, Proper in a Republic* (Philadelphia: Thomas Dobson, 1786), 30. Benjamin Rush, *An Account of the Bilious Remitting Yellow Fever* (Philadelphia: Thomas Dobson, 1794), 167. The medical roots of Rush's notions about circulation and commerce are discussed in Sari Altschuler, "From Blood Vessels to Global Networks of Exchange: The Physiology of Benjamin Rush's Early Republic," *Journal of the Early Republic* 32, no. 2 (Summer 2012), 207–231.

66. See Alan Aberbach, "Samuel Latham Mitchill: A Physician in the Early Days of the Republic," *Bulletin of the New York Academy of Medicine* 40, no. 7 (July 1964), 503–509. At the dedication of the canal, Mitchill delivered a speech and ceremoniously poured water into the canal from the Rhine, the Ganges, the Nile, and several of the other major waterways of great and lasting civilizations. The absence of work on Mitchill's political thought is unfortunate. Though a Jeffersonian, he advocated for manufacturing and thus to some degree fits the mold of "pro-manufacturing republicans" as discussed by Andrew Shankman, "'A New Thing on Earth': Alexander Hamilton, Pro-Manufacturing Republicans, and the Democratization of American Political Economy," *Journal of the Early Republic* 23, no. 3 (Autumn 2003), 323–352.

67. The many troubling facets of the West Indies in early republican minds is dealt with thoroughly in Sean X. Goudie, *Creole America: The West Indies and the Formation of Literature of Culture in the New Republic* (Philadelphia: University of Pennsylvania Press, 2006); for the West Indies trade boom and the United States, Steven Shapiro, *The Culture and Commerce of the Early American Novel: Reading the Atlantic World-System* (University Park: Pennsylvania State University, 2008); and for tangled relationships among the West Indies, commerce, and slavery, see Philip Gould, *Barbaric Traffic: Commerce and Antislavery in the Eighteenth-Century Atlantic World* (Cambridge: Cambridge University Press, 2006).

68. Here I want to refer readers back to a problem in the historiography of disease first identified by Erwin Ackerknecht in his essay "Anticontagionism between 1821 and 1867," *Bulletin of the History of Medicine* 22 (1948), 117–153, and then taken

up by a number of historians—Aisenberg, *Contagion*; Richard J. Evans, *Death in Hamburg: Society and Politics in the Cholera Years 1830–1910* (Oxford: Clarendon Press, 1987); Peter Baldwin, *Contagion and the State in Modern Europe, 1830–1930* (New York: Cambridge University Press, 1999); and for the United States, Martin Pernick, "Politics, Parties, and Pestilence: Epidemic Yellow Fever in Philadelphia and the Rise of the First Party System," *William and Mary Quarterly* 29 (October 1972), 559–586. What has survived from Ackerknecht's original argument is the notion that late-eighteenth- and nineteenth-century Western health advocates facing diseases with ambiguous causes, especially cholera, favored localism (anticontagionism or miasmatism) because it accorded with their liberal, commercial political views. Localist health measures allowed them to institute social reform measures that spread the blessings of modern sanitation to the ignorant poor, and it left commerce inviolate, unlike the quarantine measures of contagionists. The tradition has not established a paradigm for understanding how these actors rationalized and justified the harmony between disease causation and their social and political views. My argument suggests that rather than crudely and self-interestedly adopting scientific ideas for social ends, investigators saw them as plausibly fitting together as a part of the apparent design of the world. See note 15 for the Introduction to this book.

CHAPTER 5

1. Elihu Hubbard Smith, *The Diary of Elihu Hubbard Smith* (Philadelphia: American Philosophical Society, 1973), October 6, 1795, 68.

2. Benjamin Rush to Julia Rush, September 13, 1793, in *The Letters of Benjamin Rush*, ed. Lyman Butterfield (Princeton, NJ: Princeton University Press, 1951) 2:663.

3. Rush to Julia, September 21, 1793, *Letters of Benjamin Rush*, 2:673.

4. Paul E. Kopperman, "'Venerate the Lancet': Benjamin Rush's Yellow Fever Therapy in Context," *Bulletin of the History of Medicine* 78, no. 3 (Fall 2004), 539–574.

5. Rush to Webster, December 29, 1797, quoted in Benjamin Spector, "Introduction" to *Supplements to the Bulletin of the History of Medicine*, ed. Henry Sigerist, No. 9 (Baltimore: Johns Hopkins University Press, 1947), 14.

6. See, for instance, College of Physicians, *Facts and Observations Relative to the Nature and Origin of the Pestilential Fever, which Prevailed in This City, in 1793, 1797, and 1798* (Philadelphia: Thomas Dobson, 1798); Academy of Medicine, *Proofs of the Origin of the Yellow Fever, in Philadelphia & Kensington, in the year 1797, from the Domestic Exhalation . . .* (Philadelphia: Thomas and Samuel Bradford, 1798).

7. Richard Hofstadter, *The Paranoid Style in American Politics: And Other Essays* (Cambridge, MA: Harvard University Press, 1964). The "paranoid style" has also met criticism for psychologizing, even though Hofstadter disavowed the psychological or clinical implications of his term. See Gordon Wood, "Conspiracy and

the Paranoid Style: Causality and Deceit in the Eighteenth Century," *William and Mary Quarterly* 39, no. 3 (1982), 401–441. Rather than looking at early Americans as being disturbed, Wood explains the paranoid style as a causal explanation, premised on a mechanistic view of society, which held that human agency, not abstract social forces, ultimately determined the paths of society. When outcomes defied expectations, eighteenth-century intellectuals could only reasonably conclude that someone had subverted the proper direction of the society in question. See also David Brion Davis, *The Slave Power Conspiracy and the Paranoid Style* (Baton Rouge: Louisiana State University Press, 1969); David Brion Davis, ed., *The Fear of Conspiracy: Images of Un-American Subversion from the Revolution to the Present* (Ithaca, NY: Cornell University Press, 1971).

8. For the material organization of the political communities, see Joanne Freeman, *Affairs of Honor: National Politics in the New Republic* (New Haven, CT: Yale University Press, 2001). Martin Pernick, "Politics, Parties, and Pestilence: Epidemic Yellow Fever in Philadelphia and the Rise of the First Party System," *William and Mary Quarterly* 29 (October 1972), 559–586. Pernick's essay has much insight to offer about the relationship between politics and science, but where he contends that politics itself structured public discourse about yellow fever, I argue that the public arena itself shaped both political and scientific discourses.

9. The classic examination of the social bonds of Enlightenment knowledge-making is Steven Shapin, *A Social History of Truth: Civility and Science in Seventeenth Century England* (Chicago: University of Chicago Press, 1995). For the orientation of this "democracy of facts" in the early republic, see Andrew Lewis, *Democracy of Facts: Natural History in the Early Republic* (Philadelphia: University of Pennsylvania Press, 2011), esp. 13–45.

10. Study of the public sphere starts with Jürgen Habermas, *The Structural Transformation of the Public Sphere: An Inquiry into a Category of Bourgeois Society*, trans. Thomas Burger and Frederick Lawrence (Boston: MIT Press, 1991). Habermas argued that the public sphere functioned as a truly rational and transparent domain of civil society only in the long eighteenth century, when men of bourgeois status used public forums and media for rational debate about matters of common concern, and by doing so effectively guided state policies. He found the deterioration of this open public sphere with the encroachments of market forces and especially corporate manipulations. The literature on the public sphere in eighteenth- and early-nineteenth-century America is vast. A brief but effective introduction can be found in John L. Brooke, "Consent, Civil Society, and the Public Sphere in the Age of Revolution and the Early American Republic," in *Beyond the Founders: New Approaches to the Political History of the Early American Republic*, ed. Jeffery L. Pasley, Andrew W. Robertson, and David Waldstreicher (Chapel Hill: University of North Carolina Press, 2004), 207–250. This chapter contends that the fear of conspiracy revealed the splintering of that public sphere earlier than Habermas would have claimed. I am much indebted to a penetrating essay by Bryan Waterman, "The Bavarian Illuminati, the Early American Novel,

and Histories of the Public Sphere," *William and Mary Quarterly* 62, no. 1 (January 2005), 9–30.

11. James Madison, "Federalist 10," in *The Federalist Papers* (New York: Signet Classics, 2003), 73.

12. Andrew Lewis, *Democracy of Facts: Natural History in the Early Republic* (Philadelphia: University of Pennsylvania Press, 2011). Lewis points out compellingly that these changes were already under way in the domain of natural history. His late-eighteenth-century early republicans were already agitating for sterner control over their natural history in the face of popular forces that insisted on the democratization of scientific inquiry. His story ends with the triumph of professional science. The basic contours of the story—Enlightenment public science succumbs to professional, institutionalized science in the nineteenth century—are familiar.

13. Madison, "Federalist 10," 73.

14. William Chalwill, *A Dissertation on the Sources of Malignant Bilious, or Yellow Fever, and Means of Preventing It* (Philadelphia: Way and Groff, 1799), 31.

15. The argument here owes much to "Counter-Enlightenment" historiography, which may well have received its most definitive statement from Theodor Adorno and Max Horkheimer, *Dialectic of Enlightenment*, trans. John Cumming (New York: Continuum, 1988 [1st ed. 1947]). They argued that the Enlightenment thinkers were fundamentally intolerant of anything that did not meet with their enlightened views, and that their attempts to control societies on the basis of reason created a dictatorship of reason. In reality, though, counter-Enlightenment thought preceded Adorno and Horkheimer, going all the way back to Enlightenment itself and the philosophers who attacked it for its narrow, controlling nature. See Graeme Garrard, *Counter-Enlightenments: From the Eighteenth Century to the Present* (New York: Routledge, 2006); Darrin McMahon, *The Enemies of Enlightenment: The French Counter-Enlightenment and the Making of Modernity* (Oxford: Oxford University Press, 2001); also see the works of Isaiah Berlin, perhaps the principal "counter-enlightenment" philosopher, of which an excellent introduction can be found in Joseph Mali and Robert Wokler, eds., *Isaiah Berlin's Counter-Enlightenment* (Philadelphia: American Philosophical Society, 2003), 1–196. In the histories of science and medicine, Michel Foucault's works stand out. See in particular *The Birth of the Clinic: An Archaeology of Medical Perception*, trans. Alan Sheridan (New York: Vintage Books, 1994), and *The Order of Things: An Archaeology of the Human Sciences*, trans. Alan Sheridan (New York: Vintage Books, 1994).

16. For the use of slavery as a metaphor in political ideology of the revolutionary generation, see Bailyn, *The Ideological Origins of the American Revolution* (Cambridge, MA: Belknap Press of Harvard University Press, 1992), 233–246.

17. Madison, "Federalist 10," 71. Of course, some scholars have argued that Madison celebrated factions as counterbalances for competing interests. See, for example, James Yoho, "Madison on the Beneficial Effects of Interest Groups: What Was Left Unsaid in *Federalist 10*," *Polity* 27, no. 4 (Summer 1995), 587–605.

18. The hysterical, paranoid tone of 1790s political wrangling comes out clearly from any general reading in the era. See Stanley Elkins and Eric McKitrick, *The Age of Federalism* (New York: Oxford University Press, 1993). For Alien and Sedition Acts, see 590–593.

19. Ebenezer Hazard to Jedidiah Morse, April 20, 1795, quoted in Henry May, *Enlightenment in America* (New York: Oxford University Press, 1978), 254. The Bavarian Illuminati conspiracy actually was started earlier by John Robison, a Scot, in his *Proofs of a Conspiracy against All the Religions and Governments of Europe* (1798), and then adopted by the federalist New England clergy. See Vernon Stauffer, *New England and the Bavarian Illuminati Conspiracy* (New York: Columbia University Press, 1918); Davis, *The Fear of Conspiracy*, 37–42; Bryan Waterman, "The Bavarian Illuminati," 9–30. Jefferson's election in 1800 exacerbated fears of a deistic plot to undermine the pious republicanism of the United States. Jon Butler, *Awash in a Sea of Faith: Christianizing the American People* (Cambridge, MA: Harvard University Press, 1992), 219–220, claims that the Bavarian Illuminati scare and Jefferson's election did more than anything else to inflame the "religious paranoia" of early republicans.

20. Donald Hickey, "America's Response to the Slave Revolt in Haiti, 1791–1806," *Journal of the Early Republic* 2, no. 4 (Winter 1982), 364, 368–369. The fear of insurrection in part caused Jefferson to move away from trade with the former French colony of Saint-Domingue in the early 1800s. For Gabriel's Rebellion and its links to the Haitian Revolution, see James Sidbury, *Ploughshares into Swords: Race, Rebellion, and Identity in Gabriel's Virginia* (Cambridge: Cambridge University Press, 1997).

21. Priestley gives his account of the scene in Joseph Priestley, *An Appeal to the Public, on the Subject of the Riots in Birmingham* (Dublin: Hillary and Barlow, 1792), 36–38. In the United States at least, the debate between Lavoisierian chemists and the phlogiston supporters never took on a paranoid tone like the yellow fever debate; instead, it remained quite cordial. After all, there were hardly any phlogiston adherents in the United States (except Priestley after he moved there), and the chemical debate did not hold thousands of lives in the balance.

22. Robert E. Schofield gives a full account of the riots and the manner in which they came about in *The Enlightened Joseph Priestley: A Study of His Life and Work from 1773 to 1804* (University Park: Pennsylvania State University Press, 2004), 263–290.

23. Joseph Priestley to Reverend Abercrombie, August 21, 1793, in *The Scientific Correspondence of Joseph Priestley* (Philadelphia: Collins Printing House, 1891), 136. The letter reads: "The spirit of bigotry nearly bordering on that of persecution being encouraged by the Court is greatly increased in this country [England], which makes it, tho' not absolutely unsafe, yet unpleasant to live in it."

24. Jenny Graham, "Joseph Priestley in America," in *Joseph Priestley: Scientist, Philosopher, and Theologian*, ed. Isabel Rivers and David L. Wykes (Oxford:

Oxford University Press, 2008), 203–230. Priestley describes his reasoning in Joseph Priestley, *Two Sermons . . . with a Preface Containing the Reasons for the Author Leaving England* (Philadelphia: Thomas Dobson, 1794), v–xxvi.

25. Joseph Priestley to Antoine-Laurent Lavoisier, June 2, 1792, *Scientific Correspondence of Joseph Priestley*, 130.

26. Jean-Pierre Poirier, *Antoine Laurent de Lavoisier* (Paris: Pygmalion, 1993), 370–387.

27. Noah Webster to George Washington, September 2, 1790 in *The Letters of Noah Webster*, ed. Harry Warfel (New York: Library Publishers, 1953), 85–86.

28. Noah Webster, *The Revolution in France, Considered in Respect to Its Progress and Effects* (New York: George Bunce, 1794), 7, 35.

29. Henry May, *The Enlightenment in America* (New York: Oxford University Press, 1976), 252–277.

30. Noah Webster, "To the People," *American Minerva*, May 1 and 2, 1796.

31. As it would turn out, Chisholm and Currie were absolutely right. The *Hanky* did import yellow fever. The story of this fateful ship is illuminated brilliantly in Billy G. Smith, *Ship of Death: A Voyage that Changed the Atlantic World* (New Haven, CT: Yale University Press, 2013).

32. Adam Hochschild, *Bury the Chains: Prophets and Rebels in the Fight to Free an Empire's Slaves* (Boston: Mariner Books, 2005), 204.

33. Colin Chisholm, *An Essay on the Malignant Pestilential Fever Introduced into the West Indian Islands from Boullam, on the Coast of Guinea, As It Appeared in 1793* (London: C. Dilly, 1795), 83.

34. Ibid., 85.

35. Ibid., 85–87, 89, 98.

36. Noah Webster, *Letters on Yellow Fever, Addressed to Dr. William Currie*, Issue no. 9 of the *Supplements to the Bulletin of the History of Medicine*, ed. Henry Sigerist and with an introduction by Benjamin Spector (Baltimore: Johns Hopkins University Press, 1947), Letter 2: 22–23.

37. Ibid., 23. The camp and jail fevers have been identified as typhus, a bacterial disease transmitted via lice, which abound in military camps and jails.

38. Webster, *Letters on Yellow Fever*, Letter 9: 42.

39. Ibid., 43.

40. Ibid., 43.

41. Ibid., 46.

42. For a bit of background on the Sierra Leone Company and the context in which it operated, see Adam Hochschild, *Bury the Chains*, 176–180, 199–212.

43. Noah Webster, *Brief History of Epidemic and Pestilential Diseases* (Hartford, CT: Hudson and Goodwin, 1799), vii–viii.

44. Smith, *Diary of Elihu Hubbard Smith*, September 20, 1795, 60.

45. William Currie and Isaac Cathrall, *Facts and Observations, Relative to the Origin, Progress and Nature of the Fever, which Prevailed in Certain Parts of the City*

and Districts of Philadelphia, in the Summer and Autumn of the Present Year (Philadelphia: William Woodward, 1802). Martin Pernick's works have most clearly linked localism and commercialism.

46. Noah Webster, "To the People," *The American Minerva*, May 1 and 2, 1796. For the place of newspapers in the early republic, see Jeffrey Pasley, *"The Tyranny of Printers": Newspaper Politics in the Early American Republic* (Charlottesville: University of Virginia Press, 2003).

47. Webster's meaning here is a bit hard to decipher. The Latin term *"copax rationis"* is undoubtedly a misspelled version of *"capax rationis,"* meaning "capable of reason" (or perhaps the transcriber misspelled the word, since Webster, a lexicographer, would be unlikely to misspell or otherwise errantly decline Latin words). He is probably intoning a phrase from Jonathan Swift, who used the term to describe human beings, who were not fundamentally rational creatures but only capable of reason. See Christopher Fox in Jonathan Swift, *Gulliver's Travels* (New York: Palgrave Macmillan, 1995), 272.

48. Noah Webster to Benjamin Rush, December 15, 1800, in *Letters of Noah Webster*, 228.

49. Noah Webster, *An American Dictionary of the English Language* (New York: S. Converse, 1828), "faction." Webster's biographers agree that the 1790s were a crucible for the intellectual, which ended with his disenchantment and retreat into conservatism. See K. Alan Snyder, *Defining Noah Webster: Mind and Morals in the Early Republic* (Lanham, MD: University Press of America, 1990), esp. 185–213; Richard Moss, *Noah Webster* (Boston: Twayne, 1984), 48–65. They have not realized the unique impact of the yellow fever debate.

50. Medical Society of the State of New York, *Report of the Committee, Appointed by the Medical Society, of the State of New-York, to Enquire into the Symptoms, Origin, Cause, and Prevention of the Pestilential Disease, that Prevailed in New-York during the Summer and Autumn of the Year 1798* (New York: From the office of the Daily Advertiser, 1799), 7–8.

51. In their affidavits, the inspectors claimed that the sick passenger suffered from a "consumptive complaint." Since the inspectors were not doctors, and no firm diagnoses were made, we should not rule out the possibility that they were mistaken and that the patients did, in fact, have yellow fever. Ibid., 28–32.

52. Samuel Latham Mitchill, *An Address to the Citizens of New-York, Who Assembled in the Brick Presbyterian Church, to Celebrate the Twenty-third Anniversary of American Independence* (New York: George F. Hopkins, 1800), 14.

53. Ibid., 14–15.

54. Ibid., 14.

55. Quotations from Rush, "On Patriotism," *Pennsylvania Journal*, October 20, 1773.

56. Rush to Anthony Wayne, May 19, 1777, *Letters of Benjamin Rush*, 1:148. The "Thirty Tyrants" or the "Thirty" was an oligarchy that ruled Athens for a short period of time just after the defeat at the hands of the Spartans.

57. The episode plays out in *Letters of Benjamin Rush*, 1:196–210.

58. Benjamin Rush to John Witherspoon, in *Letters of Benjamin Rush*, 1:36, 45. To John Morgan, January 20, 1768, 1:49–50. Hereafter, all references to Rush's letters come from this source, unless otherwise noted.

59. Rush to John Warren, October 12, 1782, 1:288–289; Rush to Richard Price, May 25, 1786, 1:390; Rush to John Coakley Lettsom, April 26, 1793, 2:635.

60. Because of its natural ingredient, quinine, cinchona bark offers both prophylactic and cure for malaria. John McNeill, *Mosquito Empires: Ecology and War in the Greater Caribbean, 1620–1914* (New York: Cambridge University Press, 2010), 74–75.

61. Rush to Julia, September 15, 1793, 2:664.

62. Rush to Julia, September 15, 1793, 2:664; September 21, 1793, 2:673; September 24, 1793, 2:678.

63. Rush to Julia, October 25, 1793, 2:726. The "French" physicians used hot baths, clysters, and mild purgatives—they were undoubtedly refugees from Saint-Domingue. McNeill discusses the Afro-Caribbean roots of bathing yellow fever patients in *Mosquito Empires*, 81–86, especially 82. Undoubtedly the French physicians took this remedy from Afro-Caribbean inhabitants, since it was not a part of the healing repertoires of the Europeans.

64. Thomas Flexner, *Doctors on Horseback: Pioneers of American Medicine* (New York: Viking, 1937), 103–104. The section on the epidemic covers 89–108. J. H. Powell, *Bring Out Your Dead: The Great Plague of Yellow Fever in Philadelphia in 1793* (Philadelphia: University of Pennsylvania Press, 1949), 207; 199–215.

65. Rush to Julia, September 30, 1793, 2:688.

66. Rush to Julia, September 15, 1793, 2:664.

67. Rush to Julia, September 23, 1793, 678; Rush to John Redman, November 5, 1793, 2:741.

68. 2 Samuel 24; 1 Chronicles 21

69. Rush to John Redman, November 5, 1793, 2:740–741.

70. Rush to Ebenezer Hazard, December 23, 1765, 1: 22.

71. Rush to Noah Webster, December 29, 1790, 1: 530.

72. Rush to John Adams, February 12, 1790, 1:531.

73. Cobbett wrote under the pseudonym "Peter Porcupine." The episode has been recounted many times. See David A. Wilson, ed., *Peter Porcupine in America: Pamphlets on Republicanism and Revolution* (Ithaca, NY: Cornell University Press, 1994), 40–41, 228–230; Marcus Daniel, *Scandal and Civility: Journalism and the Birth of American Democracy* (New York: Oxford University Press, 2009), 187–230.

74. October 16, 1797, *Letters of Benjamin Rush*, 2:794.

75. See, for example, Benjamin Rush, *Observations upon the Origin of the Malignant Bilious, or Yellow Fever in Philadelphia* (Philadelphia: Budd and Bartram, 1799), 3, 28.

76. Rush to John Adams, December 26, 1811, 2:1114–1115. See Rush's autobiography, composed in 1800, for a retrospective view of the yellow fever years:

Benjamin Rush, *Travels through Life* (Princeton, NJ: Princeton University Press, 1948), 95–96.

77. My depiction of Rush meshes well with that by Michael Meranze in *Laboratories of Virtue: Punishment, Revolution, and Authority in Philadelphia, 1760–1835* (Chapel Hill: University of North Carolina Press, 1996); and Meranze, "Introduction to Benjamin Rush," in Benjamin Rush, *Essays: Literary, Moral, and Philosophical*, ed. Michael Meranze (Schenectady, NY: Union College Press, 1988). According to Meranze, Rush exemplified the darker side of the Enlightenment. In the latter work, he describes Rush's program as a "systematic attempt to overturn custom and replace it with ideas and practices based on Rush's notion of truth" (ix).

78. Madison discusses the politics of science in Federalist 18, 37, and 47, *Federalist Papers*.

79. Benjamin Rush, *Lectures on Animal Life* (Philadelphia: Thomas Dobson, 1799), 62. A recent essay that explains how Rush's medicine influenced his politics is Sari Altschuler, "From Blood Vessels to Global Networks of Exchange: The Physiology of Benjamin Rush's Early Republic," *Journal of the Early Republic* 32, no. 2 (Summer 2012), 207–231; Simon Finger, *The Contagious City: The Politics of Public Health in Early Philadelphia* (Ithaca, NY: Cornell University Press, 2012)

80. Again, I am referring to a historiographic tradition going back to Adorno and Horkheimer, *Dialectic of Enlightenment*. See note 15.

81. William Currie, *Medical Repository* 1, no. 3 (1798), 584.

82. William Currie, *Observations on the Causes and Cure of Remitting or Bilious Fevers: To Which Is Annexed, an Abstract of the Opinions and Practice of Different Authors; and an Appendix, Exhibiting Facts and Reflections Relative to the Synochus Icteroides, or Yellow Fever* (Philadelphia: William T. Palmer, 1798), iii–iv.

83. Currie and Cathrall, *Facts and Observations*, 28–37.

84. Ibid., 32.

85. Smith, *Diary of Elihu Hubbard Smith*, October 5, 1795, 67.

CONCLUSION

1. Benjamin Rush to John Adams, September 21, 1805, in *The Letters of Benjamin Rush*, ed. Lyman Butterfield (Princeton, NJ: Princeton University Press, 1951), 2:906.

2. John Haygarth, *A Letter to Dr. Percival, on the Prevention of Infectious Fevers* (London: R. Cruttwell, 1801), 152.

3. In a letter to John Syng Dorsey, May 23, 1804, Rush wrote of William Currie, who had recently joined the Board of Health of Philadelphia: "From the moderation he exercises towards vessels from W. India ports, a suspicion might be inferred that he has changed his opinion." See *Letters of Benjamin Rush*, 2:882–883.

4. The downfall of yellow fever in Saint-Domingue is best discussed in John McNeill, *Mosquito Empires: Ecology and War in the Greater Caribbean, 1640–1914* (New York: Cambridge University Press, 2010), 236–266. We should not totally

dismiss the potential effectiveness of public health innovations, which could have prevented the importation of the yellow fever vector and virus from Cuba, itself a hotbed of yellow fever, and the United States' favorite Caribbean trading partner after 1805. See Thomas Apel, "The Rise and Fall of Yellow Fever in Philadelphia, 1793–1805," in *Nature's Entrepôt: Philadelphia's Urban Sphere and Its Environmental Thresholds*, ed. Brian Black and Michael Chiarappa (Pittsburgh: University of Pittsburgh Press, 2012).

5. Margaret Humphreys, *Yellow Fever and the South* (Baltimore: Johns Hopkins University Press, 1999).

6. Felix Pascalis Ouvière, *A Statement of the Occurrences during a Malignant Yellow Fever, in the City of New York, in the Summer and Autumnal Months of 1819* (New York: William A Mercein, 1819).

7. Henry May, for example, sees the common-sense orientation of the American Enlightenment coming to an end around 1815. See May, *The Enlightenment in America* (New York: Oxford University Press, 1976), 303–357. Theodore Dwight Bozeman, *Protestants in an Age of Science: The Baconian Ideal and Antebellum American Religious Thought* (Chapel Hill: University of North Carolina Press, 1977), suggests that common sense survived in the scientific traditions of Princeton until the Civil War. Common sense survived particularly in religious traditions in the notion of an innate moral sense; see Mark Noll, *America's God: From Jonathan Edwards to Abraham Lincoln* (New York: Oxford University Press, 2002).

8. Andrew Lewis, *A Democracy of Facts: Natural History in the Early Republic* (Philadelphia: University of Pennsylvania Press, 2011); for the influence of Alexander von Humboldt on this tradition, Aaron Sachs, *The Humboldt Current: A European Explorer and His American Disciples* (New York: Oxford University Press, 2007). More famously, the same impulse—the romantic scientific impulse—achieved its most lasting expression in the work of Charles Darwin. See Robert J. Richards, *The Romantic Conception of Life: Science and Philosophy in the Age of Goethe* (Chicago: University of Chicago Press, 2002), 511–554.

9. John Harley Warner, *Against the Spirit of System: The French Impulse in Nineteenth-Century American Medicine* (Baltimore: Johns Hopkins University Press, 1998).

10. For debates about the cause of yellow fever in the South, see Humphreys, *Yellow Fever and the South*, esp. 17–44. For cholera, see Charles Rosenberg, *The Cholera Years: The United States in 1832, 1849, and 1866* (Chicago: Chicago University Press, 1987); for anatomy and autopsy, see Michael Sappol, *A Traffic of Dead Bodies: Anatomy and Embodied Social Identity in Nineteenth-Century America* (Princeton, NJ: Princeton University Press, 2001).

Index